Transcription

Cell and Biomolecular Sciences
A series of undergraduate textbooks focusing on modular course teaching requirements.
edited by ***David P. Phoenix***, *University of Central Lancashire, UK*

Volume One
Transcription
William M. Brown and Philip M. Brown

Transcription

William M. Brown
Restoragen, Inc.
Lincoln, Nebraska, USA

and

Philip M. Brown
Cytocell Limited
Adderbury, Oxfordshire, UK

TAYLOR & FRANCIS
ALERE FLAMMAM
Founded 1798

London and New York

First published 2002 by Taylor & Francis
11 New Fetter Lane, London EC4P 4EE

Simultaneously published in the USA and Canada
by Taylor & Francis Inc
29 West 35th Street, New York, NY 10001

Taylor & Francis is an imprint of the Taylor & Francis Group

Typeset by EXPO Holdings, Malaysia
Printed and bound in Great Britain by TJ International Ltd, Padstow, Cornwall

British Library Cataloguing in Publication Data
A catalogue record for this book is available from the British Library

Library of Congress Cataloging in Publication Data
A catalog record has been requested.

ISBN 0-415-27199-1 (hbk)
ISBN 0-415-27200-9 (pbk)

Contents

Series Preface

The evolution of modular courses and the growing realisation that single texts are unable to meet the needs of students, with some sections not presenting the required depth and other superfluous material going unread, has led to the development of this series. It is intended to provide concise yet comprehensive coverage of key areas in biomolecular science, taking students from basic through to honours-level material. The books highlight key points and use course- and advanced-level questions to aid revision and understanding, whilst at the same time providing a depth of knowledge that will allow the text to be used as a work book during the course of study.

Abbreviations

A	adenine
AP-1	activator protein-1
APCs	antigen-presenting cells
Apo-B	apolipoprotein B
ATP	adenosine triphosphate
C	cytosine
Ca^{2+}	calcium ion
cAMP	cyclic adenosine monophosphate
CAT	chloramphenicol acetyl transferase (a reporter gene)
cDNA	complementary DNA
cGMP	cyclic guanosine monophosphate
CNS	central nervous system
cos	cohesive site (in λ DNA)
CRC	class I regulatory complex (of class I MHC genes)
CRE	cAMP response element
dATP	deoxyadenosine triphosphate
DHS	DNase I-hypersensitive site
DNA	deoxyribonucleic acid
DNA-PK	DNA-dependent protein kinase
E. coli	*Escherichia coli*
EGF	epidermal growth factor
Enh-A	enhancer A (of class I MHC genes)
Fe	iron
G	guanine
GAl 4	yeast transcription factor galactose 4
GDP	guanine diphosphate
GTP	guanine triphosphate
H_2O_2	hydrogen peroxide
HLA	human leukocyte antigen
HNF1	hepatocyte nuclear factor 1
hnRNA	heterogeneous nuclear RNA
HS	hypersensitive
HSE	heat shock element
HSF	heat shock factor
HSP	heat shock protein
HTH	helix-turn-helix
IPTG	isopropyl-β-thiogalacto pyranoside
IRE	interferon response element
IRSE	interferon-responsive sequence element
JAK	Janus kinase
kb	kilobase
kDa	kiloDalton
LCR	locus control region
Mg	magnesium
MHC	major histocompatibility complex
Mn	manganese
mRNA	messenger RNA

N	any of the four nucleotides A, C, G, or T
NAP-1	nucleosome assembly protein 1
NCAM	neural cell adhesion molecule
NCs	negative co-factors
NF	nuclear factor
NK	natural killer (cell)
NMR	nuclear magnetic resonance
NRE	negative regulatory element
NTRC	nitrogen regulatory protein C
O	operator
$OH^{\bullet}$	hydroxyl radical
P	promoter
PAX	paired box
PCs	positive co-factors
PIC	pre-initiation complex
Pu	purine
Py	pyrimidine
RHD	Rel homology domain
RNA	ribonucleic acid
RNAp	RNA polymerase
RNP	ribonuclear protein
rRNA	ribosomal RNA
snRNA	small nuclear RNA
Sp1	stimulatory protein 1
Stat1	signal transducer and activator of transcription
T	thymine
TAF	TBP-associated factors
TBP	TATA-binding protein
TCR	T-cell receptor
TF	transcription factor
tRNA	transfer RNA
tsp	transcriptional start point
UAS	upstream activating sequence
UV	ultraviolet
X-gal	5-bromo-4-chloro-3-indolyl-β-D-galactoside
Zn	zinc

1 Introduction to Transcription

OBJECTIVES

a) develop an understanding of the scale of the problem of controlling the genome
b) introduce the mechanisms and essential components of transcription

1. Overview

Consider the genome; it contains genes encoding all the proteins that the cell or organism will ever need. However, at a given point in a cell or an organism's life cycle, many of these proteins will be unnecessary and it would be wasteful and pointless to express them. In multi-cellular organisms, different cells in different tissues express different subsets of genes. Again, it would be wasteful and pointless - and even conceivably harmful - for a central nervous system neurone to make muscle or liver proteins. How then should the genome be regulated? Should genes be switched "off" in the ground state and then switched "on" when needed, or should genes basically be active and then be switched "off" when not needed? Does the answer to that question depend on the size of the genome, the total number of genes, the degree of specialisation of the cells and tissues of the organism, and the fraction of those genes that need to be expressed in a cell at a given time?

2. The central dogma of molecular biology

All cells contain a complete genome, that is, the DNA that encodes all the proteins a given organism can ever make. Transcription is the production of RNA from this DNA. Translation is the production of proteins from some of the transcribed RNA, specifically the messenger RNA (mRNA). Such mRNA sequences are also referred to as "transcripts". Not all proteins are required at all times; how *selective* gene expression occurs and is regulated is the subject of this book.

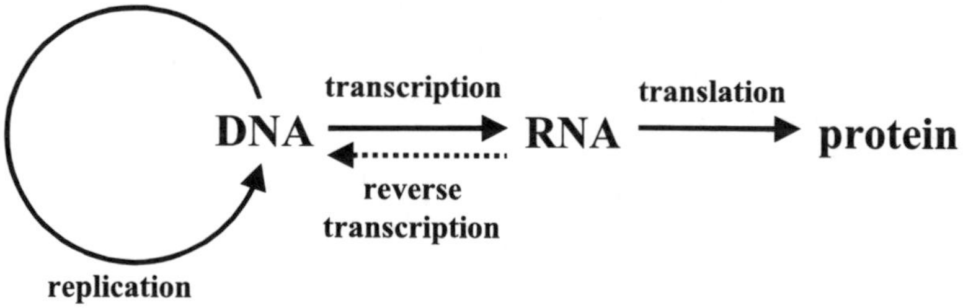

Figure 1.1 The "central dogma" of molecular biology: DNA makes messenger RNA makes protein.

2.1 How *should* transcription be regulated?

Many genes in the genome are not expressed at a given time.

The human genome comprises approximately 30,000–40,000 genes; a given cell may require expression of a small proportion of these. Thus, many or even most of the genes may not be required to be expressed at a given time in a given cell.

How then should the genome be regulated? Should all the genes be constitutively expressed, with expression of many genes being inhibited by some process or mechanism? Alternatively, should expression of all the genes be constitutively prevented, with those that are necessary being specifically activated? What are the implications of these two proposed mechanisms?

In the eukaryotic genome, most genes are "off" and are switched "on" when needed.

When phrased in this manner it seems intuitively obvious that the ground state of a gene should be "off" (*i.e.*, not expressed). Then, as and when required, a gene should be switched on for only as long as is necessary, it being extremely wasteful to produce transcripts encoding proteins that are not needed.

While this does seem logical when phrased in this way, it is not, however, a universal answer. Single cell organisms without a nucleus, such as bacteria, are termed prokaryotes. Regulation of gene expression in prokaryotes is largely *negative*, that is, the genes are "available" for expression but repressor molecules switch off genes that are not required for expression. Furthermore, in a single cell organism, with no tissue specialisation, more of the genes of the genome will be required to be expressed.

In prokaryotic genomes, most genes are "on" and are turned "off" when not needed.

It should also be obvious that if all genes were switched "off", the entire genome would be silent, with no gene being transcribed. In this case, there would be no expression of any of the proteins needed to transcribe genes, including RNA polymerase, transcription factors, splicing enzymes, ribosomal proteins or RNAs, the basic enzymes of oxidative metabolism, or anything else. Thus, on reflection it is clear that some gene products will be required by almost all cells at almost all times and perhaps these genes should be constitutively active – at some level – so that their gene products are always available in the cell. Additionally, there should also be a mechanism to activate other genes, the products of which will only be needed in some cells some of the time.

However, the basic point that most genes should be inactive most of the time is not lost. It is the most logical and economical manner to regulate the genome, and it is the mechanism evolution has adopted in higher organisms.

3. Nomenclature

Before proceeding further, this seems a good point to try to clarify the sometimes-confusing nomenclature used in discussing transcription. The following terms have generally accepted meanings that will be adopted throughout the rest of the book.

Sites on DNA are classified by their position in relation to genes and transcription start sites. Thus, the "promoter" is the region immediately in front of the transcription start site. This is, therefore, said to be "upstream of" or "5′ to" the start site (see Figure 1.2). It is on the promoter that RNA polymerase assembles. Typically, a promoter extends for around 200 base pairs upstream (*i.e.*, bases –1 to –200, there being no base zero). Generally, the eukaryotic promoter contains a TATA box and various other DNA sequences to which proteins (transcription factors) bind, as will be discussed in later chapters.

Enhancers are regions of DNA that may be located upstream or downstream or even in an intron, near the gene or distant from it and in either orientation relative to the transcription start site. Enhancers often contain multiple copies and/or several different types of binding sequences for transcription factors.

Figure 1.2 The first transcribed base is referred to as base +1; this is "downstream" of the promoter. In transcription nomenclature there is no base zero; the last base of the promoter is termed base −1, and the bases "upstream" of the transcriptional start site (sometimes referred to as the "tsp") are numbered negatively.

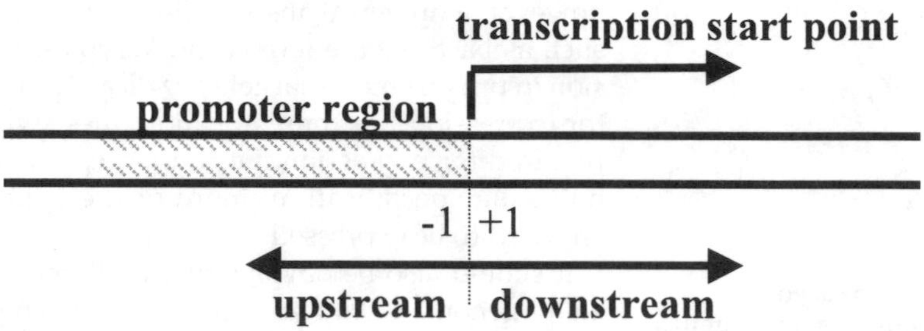

Promoters and enhancers contain DNA sequences or elements to which DNA-binding proteins bind.

DNA binding is mediated by distinct domains in transcription factors, some of which can confer DNA-binding activity on proteins that ordinarily do not bind DNA, when joined artificially to them in chimeric proteins.

The DNA sequences at which transcription factors bind are variously referred to as elements, sequences, and boxes. The proteins that bind to the sequences are variously referred to as factors, receptors (as in steroid hormone receptors), binding proteins, activators, repressors, or inhibitors.

A transcription factor must have at least two properties, that of DNA binding and activation or repression of gene expression. In addition, transcription factors need to interact with other proteins, especially those of the basal transcription machinery, including RNA polymerase. Other properties seen in transcription factors include ligand binding and nuclear localisation signals.

In many examples, it has been shown that separate domains or regions of the protein mediate the two key properties of DNA binding and gene activation. Such separation of function is most elegantly illustrated by experiments in which domains are swapped or added–using recombinant DNA techniques–to unrelated proteins. The resulting chimeric protein then exhibits the DNA-binding or activating properties of the parent protein.

Figure 1.3 Definitions

Promoter: Specific DNA sequences, usually upstream of a gene's transcriptional start site, that allow the formation of a transcription complex and allow the initiation of transcription.

Enhancer: Short DNA sequence that can increase the frequency of the initiation of transcription in a largely distance- and orientation-independent fashion.

Intron: A non-coding sequence present within a eukaryotic gene. Introns are spliced out of transcripts during processing to form mature mRNA.

Exon: Part of a eukaryotic gene that is transcribed and usually translated into a protein.

TATA box: A DNA sequence found within prokaryotic promoters, typically at approximately position −25.

For example, in the yeast transcription factor "GAL4" the properties can be experimentally separated. GAL4 is a protein that controls the transcription of the genes involved in galactose metabolism in yeast. The protein binds to a yeast enhancer known as the upstream activating sequence (UAS).

GAL4 has three functions; it binds DNA at the UAS sequence, it activates transcription, and it interacts with GAL80, another protein. GAL4 contains three domains, each responsible for one of these functions. These domains function largely independently of each other and can be joined to other proteins and protein domains to give new chimeric proteins with the predicted properties. Thus, when the DNA-binding domain of GAL4 was replaced with that from the protein LexA, the chimeric GAL4 bound only to a reporter gene containing the LexA operator sequence.

Box 1 Reporter Genes; β-galactosidase, CAT, Luciferase, and *lac-Z* Transgenic Mice

A common technique to simplify the study of transcription is that of using reporter genes attached to the DNA sequence being studied, rather than the endogenous gene to which the promoter is normally connected. A reporter gene is a gene encoding a robust enzyme with a readily quantitated activity, and removes the need for antibodies to the protein of interest or for the protein encoded by the gene to possess enzymatic activity.

For example, the bacterial enzyme β-galactosidase is widely used because it converts the colourless substance X-gal (5-bromo-4-chloro-3-indolyl-β-D-galactoside) to a blue product in a simple and versatile reaction, which is easily observed and/or quantitated spectrophotometrically. Similarly, the enzyme chloramphenicol acetyl transferase (CAT) is widely used because its protein product is robust (even in the hands of students!).

The luciferase enzyme from the firefly has also been used. This reporter gene's protein can generate light from appropriate reagents, and this can be readily measured.

In a more recent development, such reporter genes have been used in combination with transgenic mouse technology to study patterns of expression in the whole animal. The reporter gene, usually β-galactosidase, is attached to the promoter of the gene of interest and the transgene is inserted into the mouse genome using a variety of techniques. With luck, in the developing animal, the reporter gene is expressed wherever the endogenous gene would have been expressed, under the control of the same transcription factors.

4. Selective gene expression

Even a "simple" single-celled organism, such as the bacterium *Escherichia coli (E. coli)*, contains several thousand genes. The bacterium needs only a subset of these genes to survive in given circumstances while consuming a given carbon source. Additionally the genes needed to be expressed will change throughout the organism's life cycle and will depend on the bacterium's environment.

In multi-cellular organisms the control of gene expression must become more complex, simply because the genome is larger and the life cycle more complex. In the multi-cellular organism, each cell contains the same genome (the same genotype), but different cells within the organism will be specialised, will have different functions, and will require the production of a different set of proteins. Thus, different patterns of gene transcription will lead to different phenotypes in these different cells.

Under normal conditions cells tightly control transcription; it would be extremely wasteful to make unnecessary transcripts which would then unnecessarily be translated into unnecessary proteins. This high degree of control is mediated through several mechanisms and at a number of levels, including the levels of transcription, translation, and protein degradation.

4.1 Control is largely at the point of initiation of transcription

Control of transcription occurs primarily at the level of initiation.

The major control point of gene expression in all cells - eukaryotic or prokaryotic - is the initiation of transcription; it is surely the most logical and economical point too. Transcription is controlled by transcription factors that function to regulate gene expression. These proteins form specific complexes with other proteins and specific DNA sequences. Under appropriate cellular stimuli these complexes can switch expression of specific genes on or off.

4.1.1 Elements in the DNA involved in transcriptional regulation

Three types of DNA element are believed to operate in controlling the efficiency of transcriptional initiation, chromatin openers, enhancers, and promoters. Each will now be briefly discussed.

Eukaryotic DNA is organised by interaction with chromatin proteins (see chapter 7), but such interactions can lead to DNA adopting a structure that makes it difficult for RNA polymerase to interact with promoters. So-called "chromatin openers" are DNA sequences that help to unravel chromatin domains from a repressed to a potentially (more) active state, increasing the frequency of initiation by making the gene more accessible to RNA polymerase and the rest of the transcriptional machinery.

Enhancers are short DNA sequences that can increase the frequency of the initiation of transcription in a manner that is largely distance- and orientation-independent. These elements are discussed in chapter 4.

Promoters are typically bipartite in structure, there being both a core promoter and an upstream regulatory region. The core eukaryotic promoter generally comprises a TATA box, and guides RNA polymerase to the correct transcriptional start site. The regulatory region contains one or more DNA sequences to which transcription factors can bind to increase or decrease the frequency of initiation, either generally or in response to some signal. Promoters are also discussed in chapter 4.

4.2 How does an extracellular signal reach the genome?

In eukaryotes, cell phenotype is largely determined by differential gene expression. This - in turn - is controlled largely at the level of the initiation of transcription by RNA polymerase and to a lesser extent via alternative splicing events during mRNA processing and other mechanisms. Such regulation is co-ordinated by transcription factors. How then do internal and external signals regulate activity of the transcription factor and how does such a transcription factor identify the correct genes to activate?

4.2.1 *How does a transcription factor identify its target gene(s)?*

Taking the latter question first, transcription factors bind to particular DNA sequences; the genes to be activated by a given transcription factor are those containing the recognition sequence of the transcription factor in their promoters. For example, the much-studied transthyretin gene encodes a blood protein that binds thyroid hormone. Transthyretin is synthesised primarily by hepatocytes. The transthyretin gene contains ten response elements and a TATA box; six are found near the transcriptional start site and four more are located distant from it, in the so-called "distal enhancer" sequence. Transcription factors binding to these sites include C/EBP, HNF1 (hepatocyte nuclear factor 1), HNF3, HNF4 (a C_2H_2 zinc finger protein), and AP1.

4.2.2 *How does an extracellular hormone affect gene expression?*

Steroid hormones pass through cell membranes and interact with cytoplasmic receptors, causing a conformational change that leads to translocation to the nucleus where the steroid–receptor complex functions as a transcription factor.

Returning to the question of how these signals are conveyed from outside the cell to the genome, several fundamental mechanisms exist by which an extracellular message can alter the rate of initiation of transcription. For example, hormones are extracellular signalling molecules that are secreted by one cell or tissue that move around the body affecting cell function or action at a different location. Lipid-soluble hormones are those that can diffuse through cells membranes and enter cells, and include cortisol, retinoic acid, and thyroxine.

The steroid hormones are lipid-soluble and operate by passing through cell membranes, binding to cytoplasmic receptors, and activating them. This activation leads to a conformational change that causes their translocation to the nucleus, where they act as transcription factors. Steroid hormone receptors are discussed in more detail in chapter 4.

In a manner similar to that seen in GAL4 (section 3.0), the steroid hormone receptor proteins have separate domains or regions responsible for DNA binding and for hormone binding. The glucocorticoid receptor's DNA-binding domain is, for example, a cys_4 zinc finger structure (see chapter 4).

Many hormones act via cell-surface receptors, which transduce an extracellular message across the membrane to produce an intracellular effect.

However, not all hormones are lipid-soluble or activate gene transcription in this manner; instead, many hormones bind to and operate via cell-surface receptors. In this case, the "message" is passed across the membrane and channelled indirectly to the nucleus to bring about changes in gene transcription.

In broad terms, there are at least three types of cell-surface receptor involved in signal transduction in the regulation of transcription: (1) those operating via G-proteins, (2) those that are cytokine receptors and possess cytoplasmic tyrosine kinase activity, and (3) those with other enzymatic activities. These will now briefly be addressed.

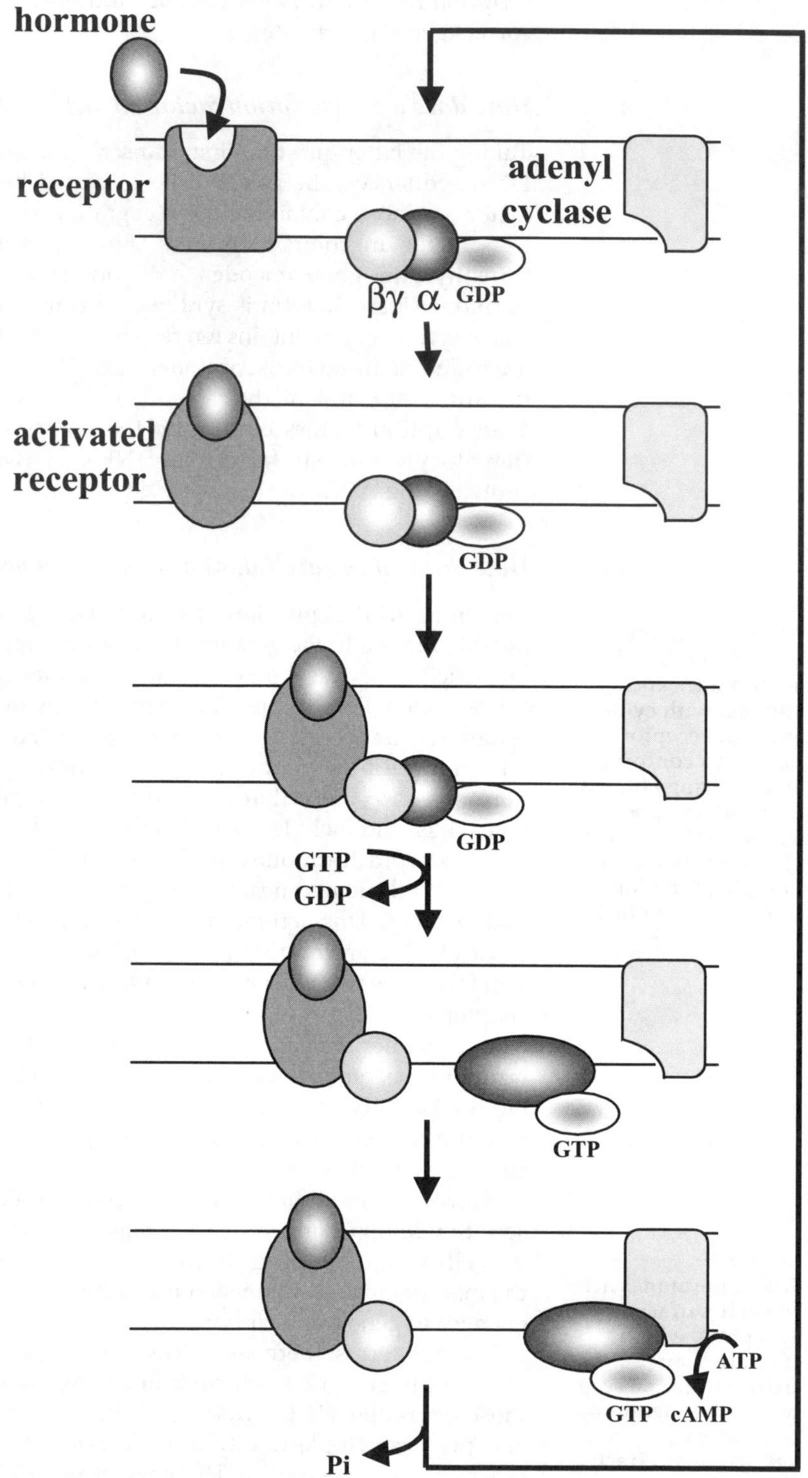

Figure 1.4 Mechanism of operation of G-protein-linked receptors. The G-protein mediates the transduction of the hormone signal from its receptor to adenyl cyclase, an enzyme that produces cyclic AMP (cAMP). cAMP – in combination with a transcription factor – can alter gene expression via interaction with DNA sequences in the promoters of various genes.

4.2.3 *Cell-surface receptors operating via G-proteins*

Second messengers such as cyclic adenosine monophosphate (cAMP) or a coupled ion channel are often associated with G-protein-linked receptors. Adrenaline (epinephrine) and glucagon operate via such G protein-linked receptors.

The G-protein-linked receptors have seven trans-membrane amino acid stretches, as is the case in the α- and β-adrenergic receptors. The G-protein mediates the transduction of the signal from the receptor to adenyl cyclase, an enzyme that produces cyclic AMP (cAMP), a second messenger with many tissue-specific effects. With regard to transcription, cAMP – in combination with a transcription factor – can alter gene expression via interaction with DNA sequences in the promoters of various genes. Such sequences are called cAMP response elements (CREs).

4.2.4 *Cell-surface receptors associated with cytoplasmic tyrosine kinase activity*

Other cell-surface receptors are associated with cytoplasmic tyrosine kinase activity as a way of transmitting the message from the extracellular ligand through the cell membrane. Gamma-interferon operates in this manner via the so-called JAK-STAT pathway. The interferon receptor dimerises upon binding γ-interferon. The receptor itself is not a kinase, but it associates with a tyrosine kinase, and a protein – which when activated by phosphorylation and translocated to the nucleus – operates as a transcription factor.

Box 2 The JAK-STAT pathway

> In the JAK-STAT pathway antigen-stimulated T-cells induce γ-interferon expression and it then affects gene expression in many cell types. Stat1 (signal transducer and activator of transcription) is a transcription factor essential to γ-interferon-mediated transcriptional activation. It is a 91-kDa protein that activates gene expression by binding to response elements in promoters. On stimulation by γ-interferon, Stat1 activity increases without there being any new protein synthesis, pointing to the activation of pre-existing Stat1 molecules. Specifically, Stat1 is activated by phosphorylation at a specific tyrosine residue. JAK (the "Janus kinase") is the kinase that phosphorylates Stat1.

Other receptors possess enzymatic activity, including serine/threonine kinase activity, cGMP cyclase activity, or phosphatase activity. Several hormones, including insulin and epidermal growth factor (EGF), operate via receptor tyrosine kinases. Typically, the receptor forms a homodimer that becomes active on ligand binding. Such receptors have an extracellular ligand-binding domain, a single trans-membrane α-helical sequence, and kinase activity in a cytoplasmic domain. The intracellular kinase activity phosphorylates intracellular proteins, thereby transducing the extracellular signal (the ligand) to produce an intracellular effect.

4.3 Negative transcription factors

While most of the well-characterised transcription factors (including those discussed in chapter 4) have a *stimulatory* effect on gene expression, the ground state of most genes being "off", it is now clear that transcription factors that *inhibit* gene transcription are also important.

A negatively acting factor could act either by interfering with the action of a positively acting factor (indirect repression) or by directly interfering with transcription by interacting with the basal transcriptional complex of RNA polymerase and associated factors (direct repression). Such direct repression, first identified only recently, has also been termed "active repression", because it is not mediated simply by steric hindrance. It remains unknown how interaction(s) between such a repressor and the RNA polymerase II basal or regulatory transcription machinery can prevent the formation or competency of a transcription complex at a promoter.

Indirect repression often operates by the negative factor preventing the positively acting factor binding to DNA. This can involve reorganisation of chromatin structure, blockage of the binding site in the DNA by binding of the inhibitory factor, or formation of a

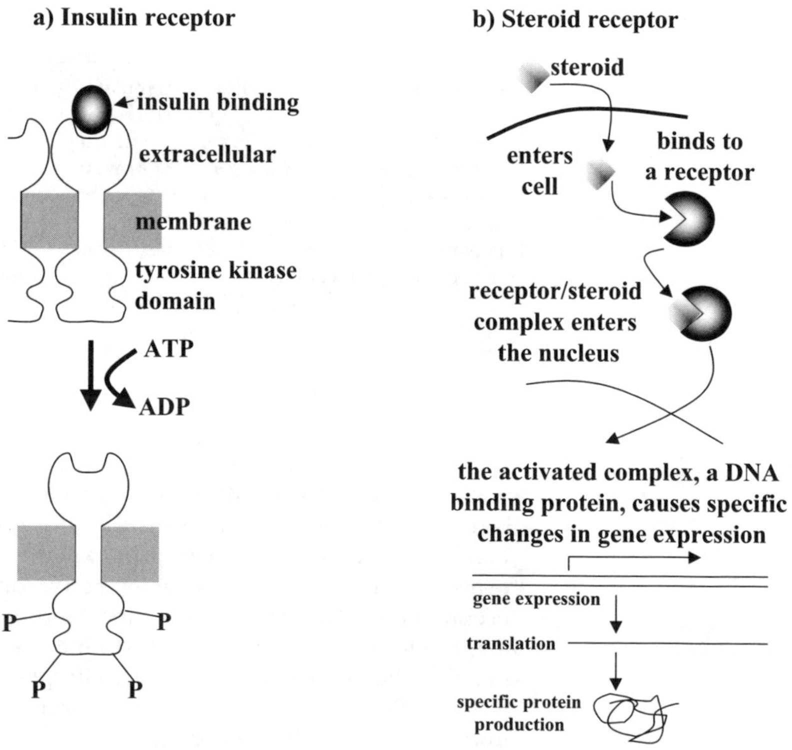

Figure 1.5 Comparison of the mechanisms of action of insulin and steroid hormones.

a) Insulin receptor. Insulin binds to an external receptor present on the cell surface. This interaction causes a conformational change, which activates the receptor's tyrosine kinase, leading to phosphorylation of an intracellular domain of the protein and of other intracellular target proteins. Thus, the extracellular hormonal signal in transduced across the cell membrane and converted into intracellular phosphorylation events.

b) Steroid hormone receptor. Steroids pass through the cell membrane, binding to an intracellular (cytoplasmic) receptor. The activated steroid–receptor complex then moves to the nucleus, where it is active as a DNA-binding protein and causes specific changes in gene expression.

A negatively acting transcription factor could simply sterically block binding of a positive factor or could interact directly with the basal transcriptional machinery.

non-DNA binding protein-protein complex. Indirect repression can also occur via quenching of the activity of a positive factor that remains bound to DNA. Such indirect repression is believed to occur in the major histocompatibility complex (MHC) class I regulatory complex (CRC), where a protein bound to the negative regulatory element (NRE) would prevent binding of positive factors to other nearby sequences in the DNA. The MHC CRC is discussed in more detail in chapter 4.

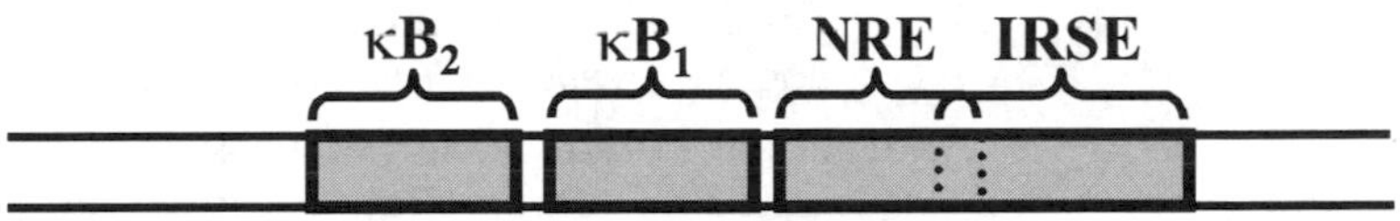

Figure 1.6 The major histocompatibility complex (MHC) class I regulatory complex (CRC) contains several closely adjacent and overlapping DNA sequences to which transcription factors bind. NRE = negative regulatory element. IRSE = interferon-responsive sequence element.

5. Other mechanisms involved in transcriptional regulation

Transcription factors are not the only mechanism involved in switching genes on and off. For example, gene methylation and histones also play an important role in gene (in)activation. Gene methylation will briefly be discussed here; the role of histones and nucleosomes is discussed in chapter 7.

Methyl groups added to 5′-CG sequences suppress transcription by altering DNA-protein interactions and, as a result, affecting the ability of transcription factors to bind promoter sequences. DNA methylation is frequently associated with gene silencing. Specialised nucleosomal structures assemble only on methylated DNA. Methylation of DNA inhibits transcription by RNA polymerase II, but appears to have no effect on RNA polymerase I transcription; it decreases transcription from some RNA polymerase III genes, but not others.

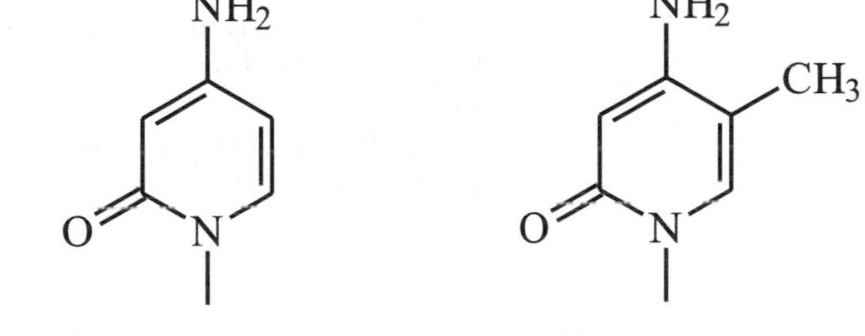

Figure 1.7 Chemical structures of
(a) cytosine and
(b) 5-methylcytosine.

SUMMARY

The genome comprises all the genes an organism can ever express. At a given point in the organism's life cycle, few of the genes may actually need to be expressed. In eukaryotes, the ground state of most genes is "off" and they are switched on as and when required. In prokaryotes, negative regulation is more common; many prokaryotic genes are active to some degree and are switched off unless needed. This switching on and off is accomplished by transcription factors. There are several fundamental mechanisms by which external signals are transduced across the cell membrane and reach the genome to alter gene expression.

FURTHER READING

1. Thieffry, D. and Sarkar, S. (1998) Forty years under the central dogma. *Trends Biochem. Sci.* **23**, 312–316.
2. Eden, S. and Cedar, H. (1994) Role of DNA methylation in the regulation of transcription. *Curr. Opin. Genet. Dev.* **4**, 255–259.
3. Suzuki, M., Brenner, S. E., Gerstein, M. and Yagi, N. (1995) DNA recognition code of transcription factors. *Protein Eng.* **8**, 319–328.
4. Latchman, D. S. (1996) Inhibitory transcription factors. *Int. J. Biochem. Cell Biol.* **28**, 965–974.
5. Finidori, J. and Kelly, P. A. (1995) Cytokine receptor signalling through two novel families of transducer molecules: Janus kinases, and signal transducers and activators of transcription. *J. Endocrinol.* **147**, 11–23.
6. Beato, M. and Sanchez-Pacheco, A. (1996) Interaction of steroid hormone receptors with the transcription initiation complex. *Endocr. Rev.* **17**, 587–609.
7. Fraser, P. and Grosveld, F. (1998) Locus control regions, chromatin activation and transcription. *Curr. Opin. Cell Biol.* **10**, 361–365.
8. Lania, L., De Luca, P. and Majello, B. (1997) Negative and positive transcriptional control during cell proliferation. *Int. J. Oncol.* **11**, 359–363.
9. Rosen, J., Day, A., Jones, T. K., Jones, E. T. T., Nadzan, A. M. and Stein, R. B. (1995) Intracellular receptors and signal transducers and activators of transcription superfamilies: Novel targets for small-molecule drug discovery. *J. Med. Chem.* **38**, 4855–4874.
10. Wasylyk, B. (1988) Enhancers and transcription factors in the control of gene expression. *Biochim. Biophys. Acta, Ser. Gene Struct. Expr.* **951**, 17–35.

END OF UNIT QUESTIONS

1. List the ways in which an extracellular hormone signal that interacts with a cell-surface receptor can lead to alterations in gene expression.
2. Explain how a steroid hormone affects gene expression.
3. Explain two generic mechanisms by which a negative transcription factor might operate.
4. Why is transcription regulated primarily at the point of initiation?

PROBLEMS

1. Compare and contrast the manner in which steroid hormones operate with the way insulin operates to affect gene expression (see reference 6).
2. Discuss the mechanisms by which negative or inhibitory transcription factors operate (see reference 4).

2 Transcription, The Process

Objectives

a) introduce the three steps in the transcriptional process
b) introduce other processes that occur co-transcriptionally or post-transcriptionally

1. Overview

The process of transcription can be thought of as a three-step process: initiation, elongation, and termination of the RNA transcript. Logically and economically, the major control point ought to be at the first step of the process, that of initiation. It is therefore not surprising that this is exactly where the cell primarily regulates transcription, but it is not the only point. In the eukaryotic cell, further levels of control are possible via mRNA processing (capping, tailing, and splicing), mRNA transport (out of the nucleus to the ribosomes), mRNA degradation, and protein translation at the ribosome.

2. The transcriptional process

An operon is a set of functionally related genes under a common control mechanism, commonly found in bacteria.

There are three primary steps in the process of transcription: initiation, elongation, and termination of the RNA transcript. Initiation involves assembling a multi-protein complex on a gene's promoter, upstream of the transcriptional start site. Elongation involves the complex moving down the DNA strand, "copying"–by base pairing–the DNA into RNA. Termination involves the complex recognising the end of the gene or operon and disassembling.

3. Post-transcriptional processing of eukaryotic mRNA

Transcription is a three-step process: initiation, elongation, and termination.

In eukaryotes, in addition to the transcription process itself, there is much processing of the nascent RNA transcript (sometimes referred to as hnRNA, heterogeneous nuclear RNA) before it passes to the cytoplasm for translation at the ribosome. Eukaryotic mRNA has a long half-life, is monocistronic (*i.e.*, each transcript is derived from one gene; operons are not found in eukaryotes), and is always modified or processed. Non-coding sequences, known as introns, that were copied from the gene have to be removed, the mRNA is "capped" with a non-transcribed structure, and the mRNA is "tailed"; that is, a poly(A) tail is added post-transcriptionally.

Eukaryotic RNA undergoes considerable post-transcriptional processing before the mature mRNA is translocated to the cytoplasm for translation at the ribosome.

Such post-transcriptional processing of the primary transcript occurs in the nucleus. It was once considered that this processing could take place anywhere in the nucleus; however, it is becoming increasingly clear that the two steps - mRNA transcription and processing - are closely linked in both time and space. Cytological studies suggest that transcription only occurs in specific places in the nucleus and that processing takes place either at the same spot or nearby. Additionally, it seems that RNA polymerase II may be involved in the splicing (removal) of introns from the nascent transcript, in addition to its role in transcription.

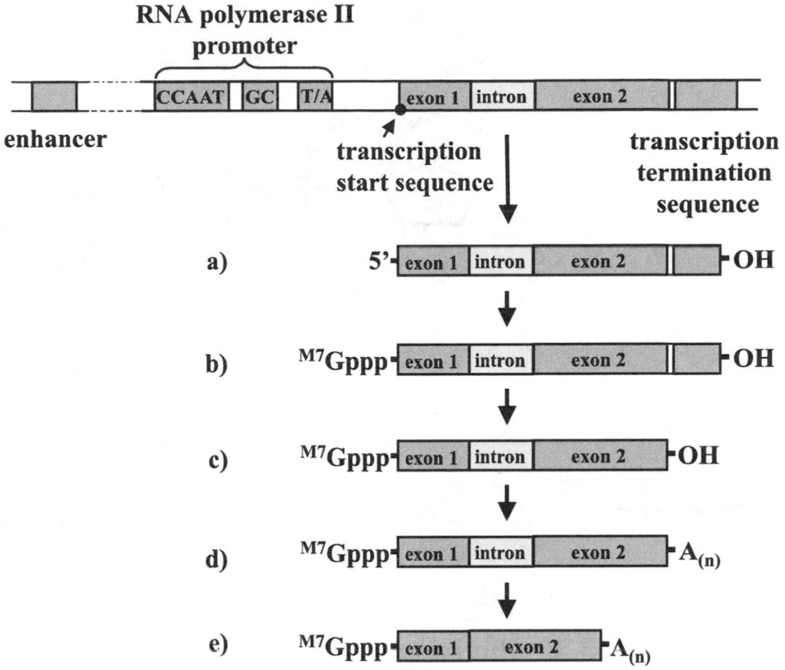

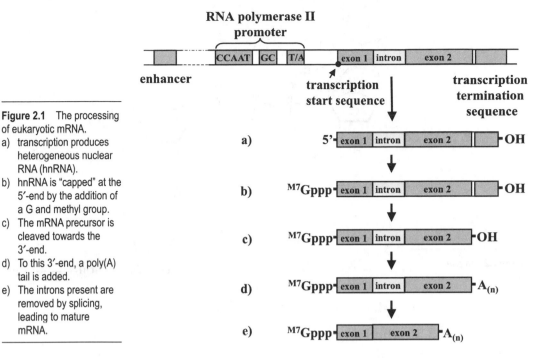

Figure 2.1 The processing of eukaryotic mRNA.

a) transcription produces heterogeneous nuclear RNA (hnRNA).
b) hnRNA is "capped" at the 5′-end by the addition of a G and methyl group.
c) The mRNA precursor is cleaved towards the 3′-end.
d) To this 3′-end, a poly(A) tail is added.
e) The introns present are removed by splicing, leading to mature mRNA.

3.1 Capping

The key eukaryotic hnRNA processing events are the G capping of the 5′-end of pre-mRNA, cleavage and addition of the poly (A) tail to the 3′-end, and splicing of introns. These processes occur in the nucleus, co-transcriptionally or post-transcriptionally.

A methylated G "cap" is added to the 5′-end of the transcript. The enzyme guanylyl transferase adds GTP in the unusual 5′ to 5′ direction.

Additionally, depending on the eukaryotic organism in question, further capping may occur. Guanine-7-methyl transferase catalyses the addition of a methyl group to the G-7 position (see Figure 2.2). 2′-O-methyl transferase likewise adds a methyl group to the 2′-hydroxyl group of the ribose moiety of the first nucleotide in the original transcript. In some transcripts, the second nucleotide of the original transcript is similarly methylated.

3.2 Tailing

Eukaryotic transcripts are cleaved at the 3′-end, downstream of a strongly conserved polyadenylation signal, given by the sequence 5′-AAUAAA. Several proteins are involved in this process. Such cleavage means that the termination of transcription does not have to be too precise. The enzyme poly(A) polymerase then tails the transcript, by adding approximately 200 A residues.

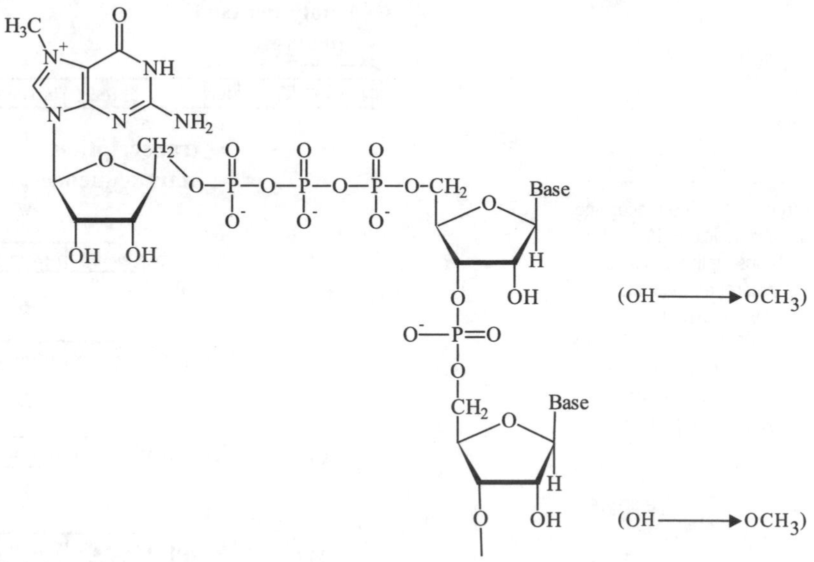

Figure 2.2 mRNA capping.

3.3 Splicing

Most eukaryotic genes are interrupted by introns or intervening sequences; introns are removed from the primary transcript in the nucleus in a process called splicing.

The "coding" sequence of most eukaryotic genes is interrupted by non-coding sequences, known as introns or intervening sequences. These non-coding sequences are removed from the transcript in a process called RNA splicing; this occurs in the nucleus. At the end of the process, the exon sequences are joined and the transcript is ready for translocation to the cytoplasm for translation.

Exons are those sequences in genomic DNA that are present in the mature mRNA. It is important not to *necessarily* equate exonic

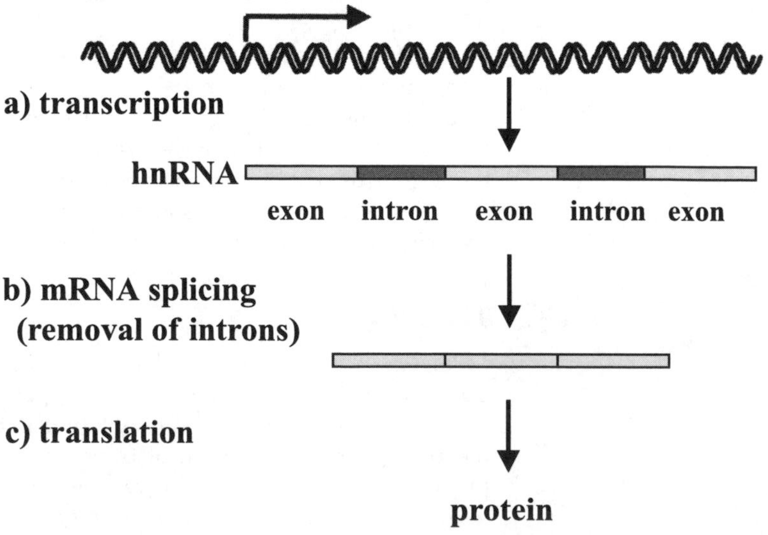

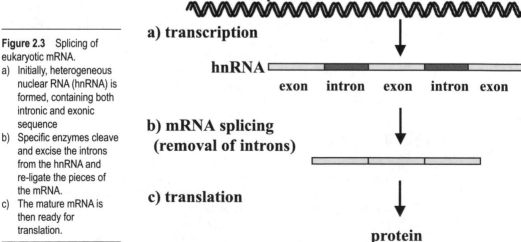

Figure 2.3 Splicing of eukaryotic mRNA.
a) Initially, heterogeneous nuclear RNA (hnRNA) is formed, containing both intronic and exonic sequence
b) Specific enzymes cleave and excise the introns from the hnRNA and re-ligate the pieces of the mRNA.
c) The mature mRNA is then ready for translation.

sequence with the sequence that will actually be translated into protein. Even exonic sequences that do end up in the mRNA may not all be translated. For example, at both ends of the mRNA there will be untranslated sequences.

The initial RNA transcribed from the chromosomal DNA contains introns, which are removed in a processing step to produce the final mRNA which is then transferred to the cytoplasm where it is translated. This removal of introns is referred to as splicing. Exons range from a few bases to a few hundred bases, whereas introns range in size from a few bases to several thousand bases. For example, Figure 2.4 lists the sizes of the introns and exons in the human *tau* gene.

The role(s) of intronic DNA and the large regions of DNA between genes in eukaryotic genomes remain largely unknown. It is perhaps non-functional and is sometimes referred to as "junk" DNA. However, it may be more sensible to be a little circumspect and simply say that we do not yet understand its role rather than to dismiss it as "junk".

Exon	Exon size (bp)	Intron size (kb)
-1	> 130	6.5
1	150	5
2	87	2.6
3	87	4.2
4	66	10
4A	753	2
5	56	1
6	198	1
7	127	3
8	54	3.5
9	266	13
10	93	> 13
11	82	1.5
12	113	10
13	208	1.3
14	> 310	

Figure 2.4 The sizes of the introns and exons of the human *tau* (τ) gene.

The "C-value paradox" is that total genome size is apparently not related to the complexity of the organism. Some amphibians have larger genomes than do humans.

Consistent with it being - largely - non-functional is the so-called "C-value paradox". The total size in base pairs of an organism's genome is referred to as its C-value. The paradox is that the C-value does not always correlate with the complexity of the organism. For example, amphibians and some plants have much more DNA in their genomes than do humans. What then can be its use?

Eukaryotic genes are expressed in monocistronic units, that is, one mRNA for each gene, in contrast to the operons seen in prokaryotic genomes (several related genes regulated and transcribed as a unit; see chapters 9 and 10). Despite this, some eukaryotic genes can give rise to more than one protein as a result of alternative splicing and/or use of more than one transcriptional start site.

As an example of an eukaryotic gene, consider the human *tau* (τ) gene. Tau, a microtubule-associated protein, binds to microtubules

In some eukaryotic genes, multiple transcripts are generated by "alternative splicing", the selective inclusion or removal of some exons.

and other cytoskeletal elements. Tau is localised to the axon of peripheral and central neurones and can organise microtubules into bundles when introduced into non-neuronal cells. Aberrant forms of the tau protein are found within neurofibrillary tangles, characteristic pathological structures in the brains of patients with Alzheimer's disease. As a result, the protein and its gene have been heavily studied.

The *tau* gene is present as a single-copy gene in the human, rat, and cow genomes; the *tau* primary transcript undergoes complex

Box 3 Splicing and the "GU-AG" rule

Analysis of many intron/exon boundary sequences had produced the so-called "GU-AG rule." This rule states that the sequence 5′-GU always occurs at the 5′-end of the intron creating the "donor" site, 5′-AG-↓-GUAAGU (the arrow indicating the bond cleaved in splicing), and the sequence 5′-AG always occurs at 3′-end of the intron, creating the acceptor site, which has the consensus structure 5′-(Py ... Py)$_{12}$-NCAG-↓-N, where "Py" means a pyrimidine, "N" is any nucleotide, and the arrow "↓" indicates the cleavage site. The following table shows the 5′ and 3′ splice sites of the introns of the human *tau* gene.

Exon	3′ splice site **cag** G (consensus)	Exon size (bp)	5′ splice site CAG **gt**aagt (consensus)
-1	N/A	>130	ACTATCAG **gt**aagcgccg Untranslated
1	Ctttcccc**ag** GTGAACTT Untranslated	150	CTG AAA G **gt**tagtggac *lys lys g*
2	Tgtgttcc**ag** AA TCT CCC *lu ser pro*	87	GCG GAA G **gt**gggccccc *ala glu a*
3	Tggtttct**ag** AT GTG ACA *sp val thr*	87	ACC ACA G **gt**gagggtaa *thr thr a*
4	Catacacc**ag** CT GAA GAA *la glu glu*	66	ACC CAA G **gt**cagtgaac *thr gln g*
4a	Ctccacac**ag** AG CCT GAA *lu pro glu*	753	CTC AAA G **gt**ctgtgtct *leu lys a*
5	Attttatc**ag** CT CGC ATG *la arg met*	56	GCC AAG **gt**aagctgac *ala lys*
6	Tatgttta**ag** ACA TCC *thr ser*	198	CTC AAG **gt**aaggaaac *leu lys*
7	Tcatttac**ag** GGG GCT *gly ala*	127	AGC TCT G **gt**aagaagaa *ser ser a*
8	Tctcttta**ag** CG ACT AAG *la thr lys*	54	GAG AGA G **gt**actcggaa *glu lys g*
9	Tccttccc**ag** GT GAA CCT *ly glu pro*	266	GGG AAG **gt**gagagtgg *gly lys*
10	Gctaccaa**ag** GTG CAG *val gln*	93	GGC AGT **gt**gagtacct *gly ser*
11	Tcatctcc**ag** GTG CAA *val gln*	82	AAA CCA G **gt**agccctgt *lys pro g*
12	Tgtgttct**ag** GA GGT GGC *ly gly gly*	113	AAA AAG **gt**aaaggggg *lys lys*
13	Cttcttgc**ag** ATT GAA *ile glu*	208	AAG CAG G **gt**ttgtgatc *lys gln g*
14	Tgctccac**ag** AA ACC CTG *lu thr leu*	>310	N/A

The intron/exon junctions of the human *tau* (τ) gene, illustrating the "GU-AG" rule. For coding regions, the protein sequence is indicated underneath the DNA sequence, in 3-letter code.

alternative splicing to generate several mRNA species of sizes between approximately 6 and 9 kb. These mRNA encode different protein isoforms, from 58 to 66 kDa. Additionally, in some neural tissues a 110 kDa isoform is observed.

The chemical mechanism of splicing involves two transesterification reactions (Figure 2.5). As a result, the exon sequences are joined and the intron is excised, either as linear RNA or as a branched "lariat" structure. The bond at the donor site is cut, and the 5′-end becomes covalently attached to a branch site in the intron in a 5′–2′ phosphodiester linking, resulting in the lariat (loop) intermediate. The branch site is typically found approximately 30 nucleotides upstream of the acceptor site and the following consensus sequence: 5′-Py_{80}-N-Py_{80}-Py_{87}-Pu_{75}-A_{100}-Py_{95} (the subscripted numbers representing the percentage frequency of that nucleotide in that position), for example, CACUGAC or UACUAAC, "Py" being a pyrimidine and "Pu" being a purine. Next, the bond at the acceptor site is cut, the exons are joined, and the intron is released as the lariat, which is then degraded.

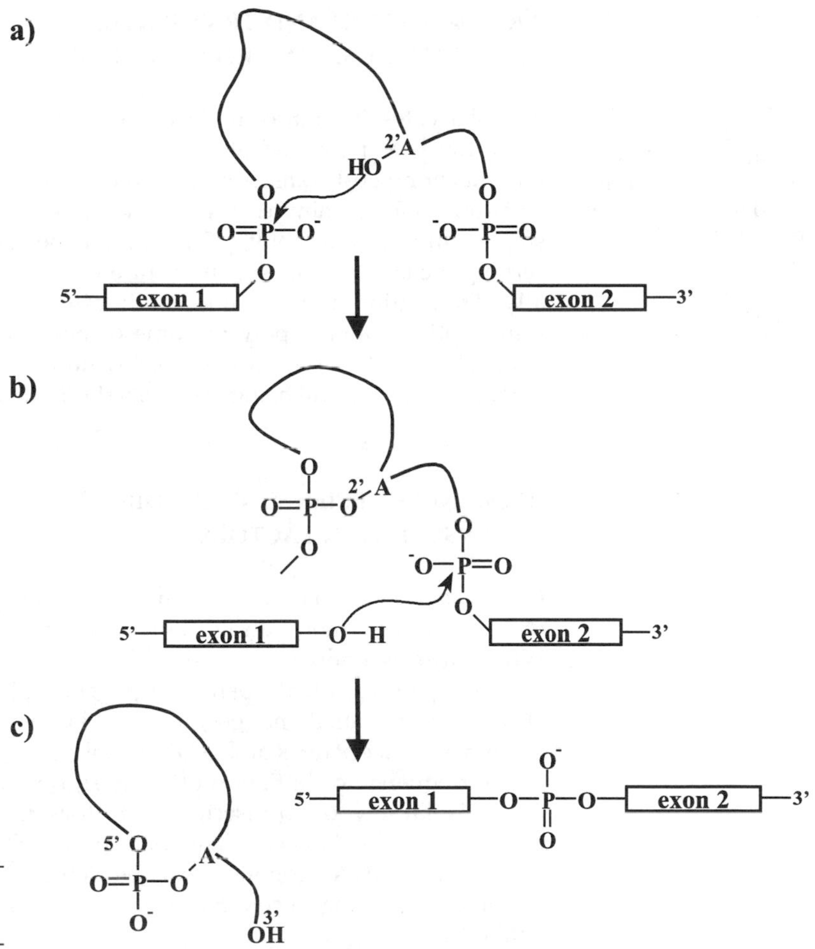

Figure 2.5 Splicing mechanism.

Splicing occurs in a large complex known as the spliceosome, which comprises approximately 40 splicing factor proteins and several small nuclear RNAs (snRNAs) which are found in complexes with nucleoproteins. Such ribonucleoprotein (RNP) complexes are referred to as snRNP particles. The snRNPs are known as U1, U2, U5, and U4/U6, and contain the U1, U2, U4, U5, and U6 snRNAs; they are strongly conserved among eukaryotes.

3.4 Editing

In some cases in addition to the three primary RNA modifications, capping, tailing, and splicing, editing of the RNA occurs, that is, insertion, deletion, or substitution of nucleotides. While it does occur in some human genes, for example the apolipoprotein B (*Apo-B*) gene (where the transcribed "CAA" is post-transcriptionally edited to "UAA", a stop codon), it is more common in lower eukaryotes.

4. Transcription and translation occur simultaneously in prokaryotes

In prokaryotes, post-transcriptional processing is rare and translation begins as soon as the mRNA transcript emerges from the RNA polymerase.

In prokaryotes, in contrast to the eukaryotic systems described above, the mRNA is translated almost as soon as it is transcribed, with minimal or no post-transcriptional processing. Prokaryotic mRNA has a half-life of a few minutes, is frequently polycistronic (meaning that a single transcript is a "copy" of more than one gene; see chapter 9 concerning the lactose operon), and is rarely modified post-transcriptionally. Translation occurs immediately after or simultaneously with transcription. Often a polyribosome or polysome results; this is the name given to the structure comprising multiple ribosomes present on a transcript that is still being transcribed (Figure 2.6).

5. Sequence-specific DNA-binding proteins or "transcription factors"

Cells respond to signals by switching on or off expression of given sets of genes and by modulating the level of transcription from genes. While little is known about the high-level mechanisms and biochemical pathways that operate to integrate physiological signals to effect transcriptional changes, much is known about the molecular machinery that operates on the DNA molecule.

The frequency of initiation of transcription and mRNA synthesis depends largely on transcription factors that bind elements in promoters and enhancers. Sequence-specific DNA-binding proteins that bind to DNA sequences can be detected in cell extracts and identified according to conserved structural features or by the use of antibodies.

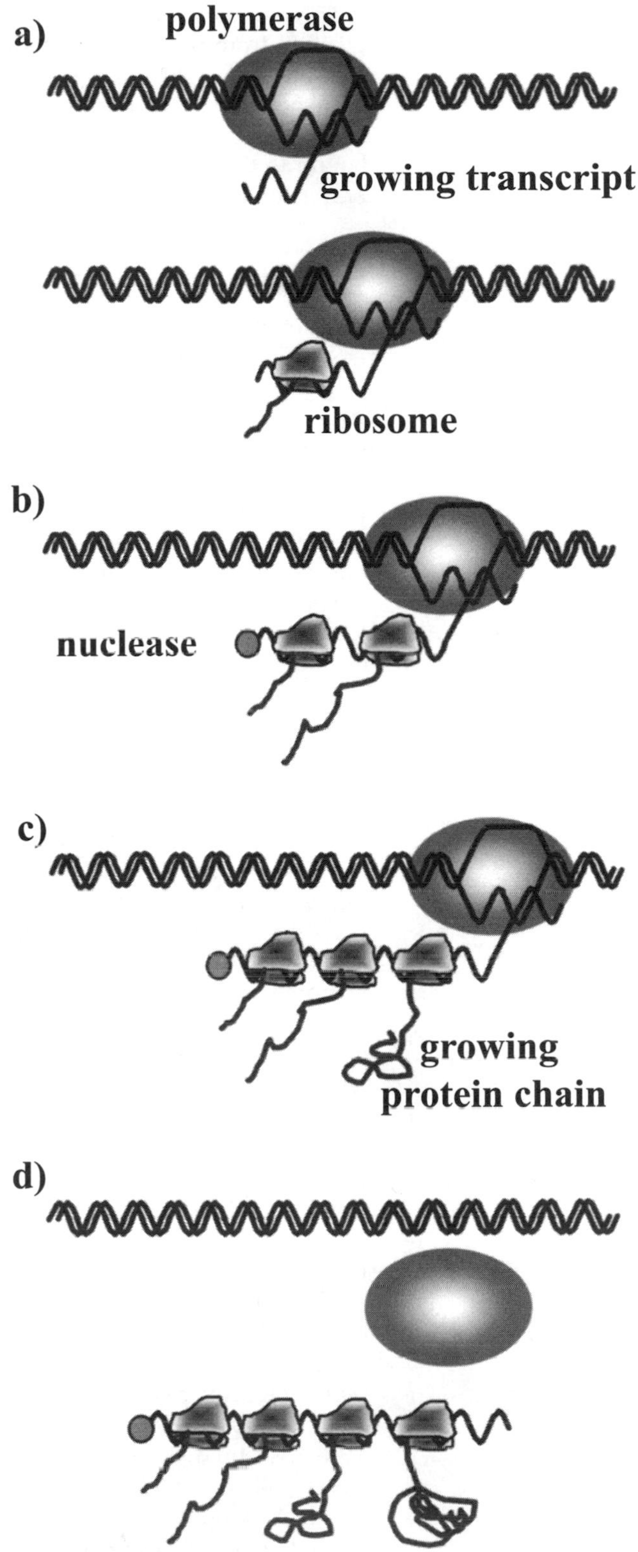

Figure 2.6 The polysome; transcription, translation, and mRNA degradation all occur *simultaneously* in bacteria.

a) As mRNA is transcribed, ribosomes bind and begin to translate it.
b) As the length of the mRNA increases more ribosomes bind. However, the window of translation is limited because the RNA will be degraded by cellular nucleases.
c) The degradation is slower than transcription or translation and is likely to occur by discrete cuts in the mRNA.
d) The completed mRNA is released; translation and degradation continue.

These proteins are present in all cells at low levels and can be purified from nuclear extracts using, for example, sequence-specific DNA-affinity chromatography. In this technique, short DNA sequences can be chemically joined to chromatography beads, nuclear extract can be passed through a column of the material, and proteins bound to the DNA sequences can be eluted by changing the salt and buffer conditions.

5.1 DNA-binding domains

Many transcription factors share conserved DNA-binding domains.

Many mammalian transcription factors have now been studied; their DNA-binding properties have been localised to relatively small subregions or domains within the proteins. These domains have been strongly conserved through evolution and are found in many DNA-binding proteins.

These domains facilitate DNA binding via a mechanism that is largely independent of the rest of the protein and can confer DNA-binding abilities on heterologous proteins. Such DNA-binding domains are necessary but not sufficient for transcriptional activation.

Three well-characterised DNA-binding motifs are the zinc finger, the homeodomain, and the C/EBP/leucine zipper motif. Many other examples outside these three classes are also known.

5.1.1 The zinc finger

The zinc finger motif was first identified in the transcription factor TFIIIA from *Xenopus laevis*, the South African clawed toad. TFIIIA is necessary for efficient transcription from the *Xenopus* 5S rRNA genes by RNA polymerase III. This 40 kDa protein binds to a 50-base-pair region located within the sequence of 5S rRNA genes.

TFIIIA comprises nine repeating units of approximately 30 amino acids. Each unit contains two conserved cysteine and two conserved histidine residues, as well as several other conserved amino acids. In the original zinc finger model, the cysteine and histidine pairs formed a tetrahedral co-ordination site for a single zinc (Zn^{2+}) ion, and the amino acids between these co-ordination sites project out as "fingers" which interact with DNA (Figure 2.7).

More recently, a second model has been proposed in which an antiparallel β-sheet and α-helix is formed between the zinc co-ordination sites. A similar structure has been observed in some other metalloproteins and the zinc finger could bind the metal ion in an analogous manner (Figure 2.8).

TFIIIA has been extensively studied; in addition to the zinc-finger-containing amino-terminal region (approximately 280 amino acids), which binds both DNA and RNA, the protein's carboxyl-terminal region (approximately 65 amino acids) is involved in the transactivation process by interacting with other general transcription factors. TFIIIA is a unique transcription factor in the sense that it binds to two structurally different nucleic acids, DNA and RNA.

From sequence analysis, many proteins contain regions that strongly resemble the zinc finger motif of TFIIIA. These proteins are

Box 4 Gel Shift Assays

The gel shift assay is used to study DNA binding by proteins. If a piece of DNA has a given electrophoretic mobility, this will be altered (the band will be shifted on the gel) if a protein binds to the DNA.

Gel shift assays are used to study DNA-protein complexes; the basic concept is that a piece of DNA has a given mobility in non-denaturing gel electrophoresis and that this mobility will be different if a protein is bound to the DNA. Thus, it is possible to detect DNA binding by a protein by the altered electrophoretic mobility of the DNA. Clearly "mild", non-denaturing conditions are essential if the protein is to retain its DNA-binding structure. Gel shift assays are variously referred to as electrophoretic mobility shift assays (EMSA) and gel retardation assays.

Specificity of the identified interaction can be demonstrated by adding related unlabelled DNA sequences in large molar excess to compete with the radiolabelled DNA.

Extracts of nuclear proteins, likely to contain transcription factors, are typically prepared using the near-universal Dignam method, which involves harvesting cells from their growing medium, pelleting them by gentle centrifugation, washing them with cold Ca^{2+}- and Mg^{2+}-free phosphate-buffered saline, and then resuspending them in a low salt buffer. Cells are then lysed by addition of the mild detergent NP-40. Intact cell *nuclei* are then pelleted by gentle centrifugation and resuspended in a low salt buffer containing 25% (v/v) glycerol, which disrupts the nuclear membrane. Aliquots of this mix are snap-frozen in liquid nitrogen and stored at –70°C until use.

Complementary oligonucleotide sequences of interest are chemically synthesised, annealed (by heating in water to near 100°C, and cooling slowly), and labelled with [α-^{32}P]dATP, using a bacterial DNA polymerase. Radiolabelled probe sequences are then purified from the labelling reaction mix and are incubated with nuclear extracts. The binding complexes are then separated by polyacrylamide gel electrophoresis.

The DNA will have a given mobility on the electrophoresis gel; if a protein binds to it, its mobility will be retarded and the radioactive signal will be seen further up the gel, hence "gel shift" assay.

In a non-denaturing acrylamide gel, a piece of DNA has a given electrophoretic mobility (a). If when mixed with proteins, or nuclear extract containing proteins, a protein binds to the DNA, when the complex is analysed on a gel, the complex will be retarded relative to the DNA (b). Hence a "gel shift" assay.

Supershift assay. If before electrophoresis an antibody is added to the DNA–protein mix and the antibody is specific to the protein that bound to the DNA, then a ternary DNA–DNA-binding protein–antibody complex will be formed and this larger complex will have an even slower electrophoretic mobility (c). Hence a "supershift" assay.

Supershift Assay

A so-called "supershift assay" takes the gel shift assay one stage further; to reliably identify the protein bound to the DNA probe sequence, antibodies are added to the mixture. If an antibody should bind to the DNA-binding protein bound to the DNA, this larger complex will have an even slower electrophoretic mobility; hence the complex will be "supershifted" relative to the original DNA's mobility.

In each supershift experiment, a competitor control is performed. Unlabelled oligonucleotide in 100-fold molar excess is added to the reaction mix before the addition of the antiserum. Such a large excess of unlabelled probe should displace the radioactive probe (leading to the absence of a band on the autoradiograph), if the binding is specific.

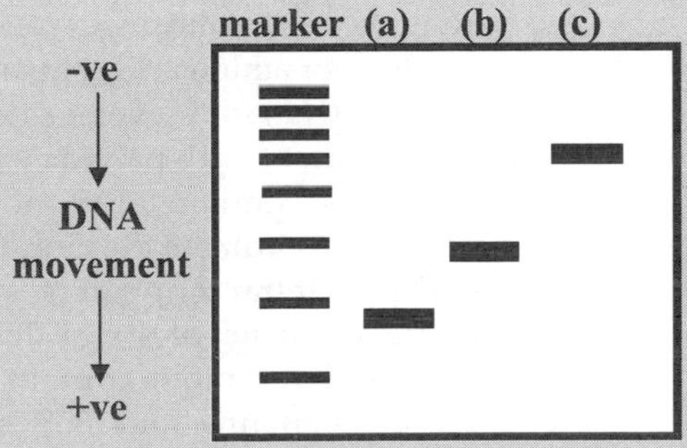

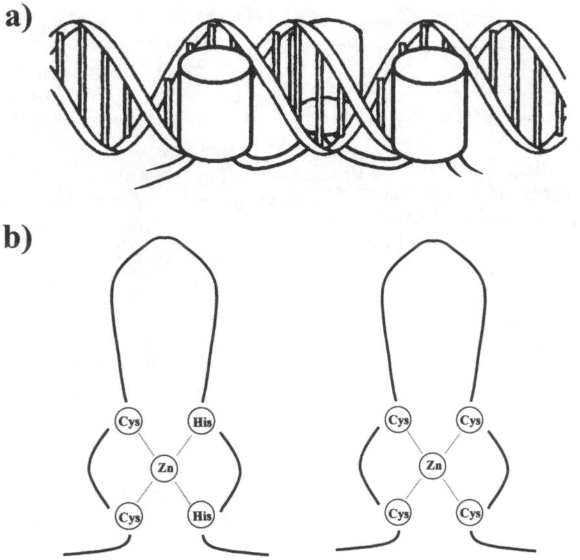

Figure 2.7 (a) Diagram showing how zinc fingers could project into the major groove of DNA. (b) Representation of the "zinc finger" structure, showing four key cysteine residues co-ordinating the zinc (Zn^{2+}) ion.

present in organisms ranging from yeast to humans. Some of these proteins are known to be transcriptional activators, while others have distinct roles in development and sex determination. The number of zinc fingers contained within a particular protein ranges from two to more then ten. In some cases, it has now been demonstrated that these proteins do bind specifically to DNA, and that both the zinc finger region of the protein and the zinc ion (Zn^{2+}) are necessary for DNA binding.

Given the strong conservation of the zinc finger sequence - and presumably structure - and the observation that differing zinc finger proteins have different DNA-binding specificities, it seems that this specificity must be determined primarily by non-finger sequences. Consistent with this, a synthetic peptide containing two zinc fingers from the yeast protein ADR1 is unable to recognise its normal target site, even though it does form a discrete structure in the presence of zinc ions that can interact non-specifically with DNA. As another example, the slightly different DNA-binding specificities of the progesterone and estrogen receptors seem to depend on differences in several non-conserved amino acids at the base of the finger region.

Analysis of amino acid sequences of zinc finger proteins shows that they readily fall into several categories. The first zinc finger identified contains a motif with paired cysteine and histidine residues, the C_2H_2 family. A second, related motif utilising (at least) four cysteine residues to coordinate the zinc ion is also known.

In the yeast transcription factor GAL4, the putative finger sequence contains a cluster of six cysteine residues (hence, the C_6 family). The steroid receptors contain two distinct finger types; the first subgroup contains four (C_4) and the second five (C_5) conserved cysteines. The C_4 family of zinc fingers, by analogy with the C_2H_2 class,

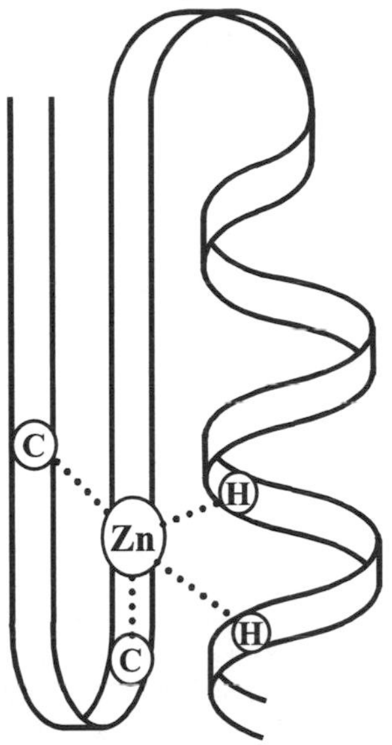

Figure 2.8 Possible alternative structure of a zinc finger.

is expected to form a similar structure, and presumably co-ordinates the zinc ion in a tetrahedral manner. The presence of more than four cysteines in some families (C_5 and C_6) at least raises the possibility that "shared" cysteine residues may co-ordinate multiple metal ions. Such an interaction is seen in metallothionein proteins, where three metal ions are co-ordinated between nine cysteine ligands.

A related structure is seen in the DNA-binding domains of steroid hormone receptors. Such zinc fingers have two pairs of cysteine residues that also tetrahedrally co-ordinate the Zn^{2+} ion. Mutagenesis experiments that remove the cysteine residues result in loss of function.

5.1.2 *The homeodomain*

The eukaryotic homeodomain DNA-binding structure is related to the helix-turn-helix structure seen in prokaryotic DNA-binding proteins.

The first and most studied structural motif in a DNA-binding domain is the helix-turn-helix motif found in many prokaryotic activator and repressor proteins. Related structures have since been identified in eukaryotic transcription factors. Using x-ray crystallographic analysis of proteins containing the motif, the structure of this DNA-binding motif has been determined.

As the name suggests, the structure is formed from two protein α-helices that are separated by a β-turn. Despite considerable sequence variability, the helix-turn-helix motifs have a highly conserved three-dimensional structure. Amino acids within one of these helices, the "recognition helix", directly contact bases in the major groove of DNA. The second helix lies across the major groove and makes non-specific contacts with the DNA (Figure 2.9).

The specificity of these proteins has been tested by swapping the helices between different proteins. This alters the binding specificity of the protein, allowing it to bind to differing target sequences. The prokaryotic helix-turn-helix proteins bind to their target sites as

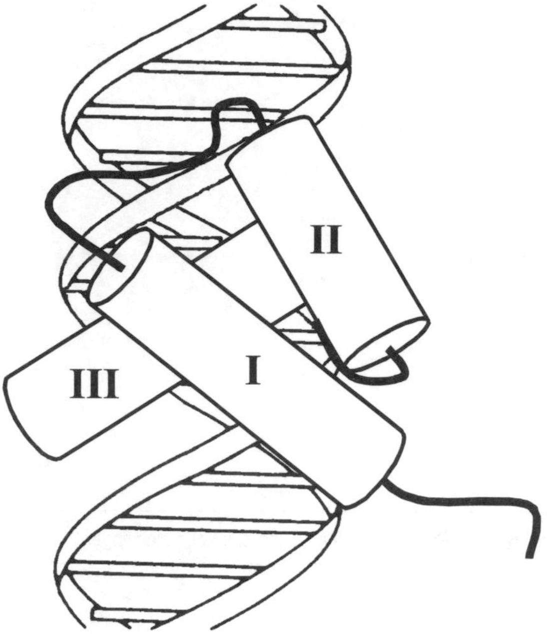

Figure 2.9 Representation of the homeodomain structure, illustrating the "helix-turn-helix", which, as its name suggests, comprises two short α-helices linked by a β-turn. The third helix lies within the major groove of the DNA and accounts for most of the protein's binding specificity.

dimers. These target sites have dyad symmetry (*i.e.*, are palindromic, "mirror images" of each other), allowing both helix-turn-helix motifs in the dimer to bind to the DNA.

In eukaryotes, helix-turn-helix motifs were first identified in the fruit fly *Drosophila*, the paradigmatic examples being *Antennapedia* and *bithorax*. The motif was identified within a family of proteins that control many of the key decisions in cellular development. These proteins were found to contain a highly conserved region, now known as the homeodomain; this sequence is 61 amino acids in length and is discussed in chapter 3. The homeodomain is rich in lysine and arginine, making it very basic. Additionally, homeodomain proteins were found to be localised to the nucleus. Within this homeodomain lies a helix-turn-helix motif similar to that found in prokaryotic proteins.

The idea that the homeobox encodes a true protein *domain* (a structural element largely independent of the rest of the protein) is supported by the manner in which the isolated homeodomain binds DNA in essentially the same fashion as the entire protein. In addition, the homeodomain is encoded by a single exon, consistent with it having been copied and moved around in genes during evolution.

Other evidence that the homeodomain binds DNA comes from many biochemical, biophysical, and *in vivo* studies. Furthermore, it has now been shown that various homeodomains do bind DNA in a sequence-specific fashion *in vitro*, and that such binding can be altered or destroyed in a largely predictable fashion by site-directed mutagenesis and "homeodomain swap" experiments. In swap experiments, the DNA-binding domain is transferred to a chimeric protein, and the resulting chimeric protein has the DNA-binding specificity of the original protein from whence the homeodomain came.

While the predicted secondary structure of the homeodomain resembled that of the helix-turn-helix motif previously characterised in prokaryotic DNA-binding proteins, several important differences are also evident. Most notably, homeodomain proteins bind to DNA as *monomers*, whereas the prokaryotic proteins bind palindromic DNA sequences as *dimers*.

The helix-turn-helix motif in homeodomains consists of three helical regions (I, II, and III) that fold into a tight globular structure. Helix I is preceded by an amino-terminal arm and is separated by a flexible loop from helix II which – with helix III – forms the helix-turn-helix (HTH) structure. Helix III provides the primary DNA recognition function lying in the major groove and establishing base-specific binding, though a homeodomain protein lacking certain side chains from helix III can still bind DNA and direct transcriptional repression.

However, comparison of these interactions with those of prokaryotic HTH domains reveals that the residues used for base interaction are displaced one helical turn towards the carboxyl-terminus. Homeodomains have an extended carboxyl-terminus where many basic amino acids are located; these can interact with DNA phosphate groups. From this, it has been proposed that the homeodomain recognition helix (helix III) does not use prokaryotic HTH-type DNA-binding geometry, but a type not seen before, which has been referred to as a "probe helix" (PH). Using the amino-terminal region of the homeodomain, additional specific contacts with bases within the minor groove are made, giving the homeodomain proteins sequence-specific recognition. In addition to the recognition helix and amino-terminal arm, the loop between helices I and II also interacts with the DNA backbone.

5.1.3 *Binding of homeodomain proteins to DNA*

The majority of homeodomain-containing proteins identified to date bind the core DNA sequence 5′-$T_{[1]}N_{[2]}A_{[3]}T_{[4]}(G/T)_{[5]}(G/A)_{[6]}$, where N is variable. However, exceptions to this are also known and some homeodomains show different sequence specificities. The specificity of homeodomain protein binding arises largely from amino acids positioned within the recognition helix (helix III) and in the amino-terminal arm. The key amino acids within these regions vary depending on the homeodomain protein in question. Within the recognition helix, nearly all homeodomains have an asparagine at position 51 and all homeodomains have a glutamine at position 50. These amino acids seem to dictate the preference for the dinucleotide at the 3′-end of the core sequence (positions 5 and 6).

The amino-terminal arm of the homeodomain is also believed to contribute to DNA-binding specificity. Residues 3 and 5 of the homeodomain region make contact with the DNA at positions 1 and 2 of the core sequence referred to above; these contacts occur within the DNA minor groove. These positions are not variable, but differences in the neighbouring residues will have differing effects on the local conformation, which may, in turn, affect specificity. For example, the *abdominal-B* homeodomain, which has lysine at position 3, prefers to bind the sequence 5′-TTAT(G/T)(G/A). In contrast, *deformed*,

Antennapedia, and *ultrabithorax*, all of which have arginine at position 3, prefer the sequence 5′-TAAT(G/T) (G/A).

The many results briefly summarised here suggest that differences in the primary structure of the homeodomains, especially within the amino-terminal arm, do cause subtle binding differences that can be demonstrated experimentally *in vitro*. These differences probably have real effects *in vivo* but given the size of the genome, they alone are likely to be insufficient to account for all the specificity and selectivity seen. The other non-homeobox regions/domains of the proteins are presumably involved in protein–protein interactions and mediating the effects on transcription.

A related structure is seen in the so-called "POU" (*Pit, Oct, Unc*) family of transcription factors; the 160-amino-acid POU domain comprises a homeodomain and a second region, the POU box, a conserved sequence that is believed to be involved in protein–protein interactions and DNA binding.

5.1.4 *C/EBP and the leucine zipper*

Another common DNA binding domain was first described in the mammalian enhancer binding protein, "C/EBP". This domain comprises a conserved stretch of approximately 30 amino acids with a substantial net basic charge, followed by a sequence containing four leucine residues at seven-amino-acid intervals. The former basic region appears to be involved in DNA binding, but not dimerisation. The latter sequence has been named the "leucine zipper" and is involved in dimerisation and DNA binding. Such dimerisation is believed to be stabilised by hydrophobic interactions between the α-helical leucine repeat regions, the leucine residue side chains intercalating in a structure resembling a zip fastener (Figure 2.10).

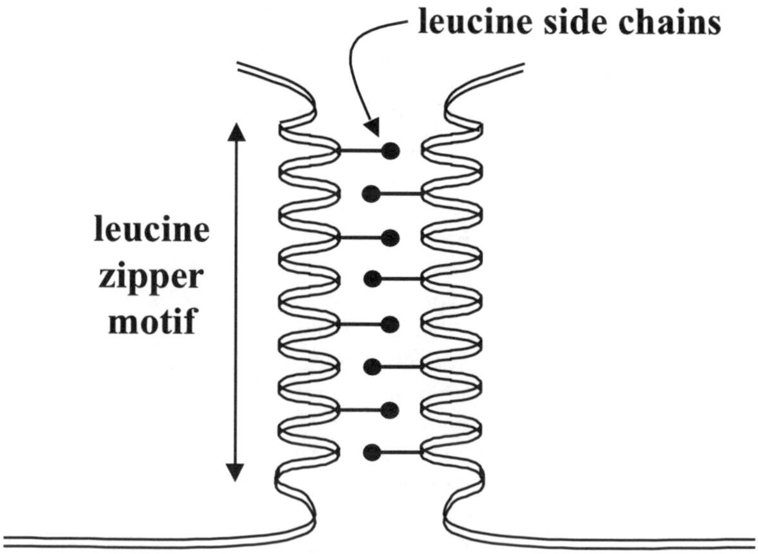

Figure 2.10 The leucine zipper dimerisation motif. The leucine side chains extend from the α-helix and interact with each other.

Proteins containing leucine zippers include c-fos and c-jun. The proteins fos and jun stimulate cells to divide and have been implicated in cancers. The proteins form homo- and heterodimers.

Summary

Transcription can be thought of as a three-step process: initiation, elongation, and termination. In prokaryotes, translation occurs simultaneously with transcription. In contrast, in eukaryotes, there is extensive post-transcriptional processing of the transcript before it moves out of the nucleus for translation. Many transcription factors fit clearly into larger families, sharing DNA-binding domains and other conserved features. These domains operate largely independently of the rest of the protein, as has been demonstrated by producing chimeric proteins using recombinant DNA techniques.

Further reading

1. Reeder, R. H. and Lang, W. H. (1997) Terminating transcription in eukaryotes: lessons learned from RNA polymerase I. *Trends Biochem. Sci.* **22**, 473–477.
2. Hertel, K. J., Lynch, K. W. and Maniatis, T. (1997) Common themes in the function of transcription and splicing enhancers. *Curr. Opin. Cell Biol.* **9**, 350–357.
3. Ranish, J. A. and Hahn, S. (1996) Transcription: basal factors and activation. *Curr. Opin. Genet. Dev.* **6**, 151–158.
4. Calkhoven, C. F. and Ab, G. (1996) Multiple steps in the regulation of transcription factor level and activity. *Biochem. J.* **317**, 329–342.
5. Andreadis, A., Brown, W. M. and Kosik, K. S. (1992) The structure and novel exons of the human tau gene. *Biochemistry* **31**, 10626–10633.
6. O'Neill, L. A. J. and Kaltschmidt, C. (1997) NF-κB: a crucial transcription factor for glial and neuronal cell function. *Trends Neurosci.* **20**, 252–258.
7. Rio, D. C. (1992) RNA processing. *Curr. Opin. Cell Biol.* **4**, 444–452.
8. Lopez, A. J. (1998) Alternative splicing of pre-mRNA: developmental consequences and mechanisms of regulation. *Annu. Rev. Genet.* **32**, 279–305.
9. McCracken, S., Fong, N., Rosonina, E., Yankulov, K., Brothers, G., Siderovski, D., Hessel, A., Foster, S., Shuman, S. and Bentley, D. L. (1997) 5′-Capping enzymes are targeted to pre-mRNA by binding to the phosphorylated carboxyl-terminal domain of RNA polymerase II. *Genes Dev.* **11**, 3306–3318.
10. Treisman, J., Harris, E., Wilson, D. and Desplan, C. (1992) The homeodomain: a new face for the helix-turn-helix? *BioEssays* **14**, 145–150.

End of Unit Questions

1. List the post-transcriptional processing steps that eukaryotic RNA transcripts may undergo before the mRNA leaves the nucleus for translation.
2. What does the "GU-AG" rule refer to?
3. How is it known that DNA binding is mediated primarily by distinct domains within transcription factors?

Problems

1. Discuss the ways in which transcription factor activity can be regulated (see reference 4).
2. Discuss RNA processing in eukaryotes (see reference 7).

3 Transcription Factors in Human Diseases

Objectives

a) demonstrate that transcription and defects in it have real-world effects and are known to cause human diseases

1. Overview

Because transcription factors are control molecules and regulate many genes, defects in transcription factors affect many genes.

Many human disorders are believed to be caused by defects in transcription, including mutations in the genes encoding the various components of the transcriptional machinery. It should be readily apparent that while a mutation in–for example–the hemoglobin gene may lead to defective hemoglobin, it is unlikely to cause effects beyond that. However, a mutation in a transcription factor that itself controls the expression of dozens of other genes is likely to have wide-ranging effects. This is indeed the case; the dramatic effects of mutations in transcription factor genes in *Drosophila*, in which antennae can be transformed into extra legs, illustrates this well (discussed below). Because of their involvement in disease and their position as control molecules, transcription factors have become a subject of investigation with respect to drug discovery. Several biotechnology companies have been expressly set up to investigate and develop transcription-related drugs.

2. Homeodomain proteins and their role in homeotic transformations

A mutation in a single homeobox gene can cause *Drosophila* to develop an extra leg instead of an antenna, or a second pair of full wings to develop in place of the halteres (balancing organs).

A developmental abnormality in which one part of the body develops in the likeness of another is referred to as a "homeotic transformation". Homeotic transformations, while known in higher animals, including humans, have been most heavily studied in *Drosophila*. The first homeotic genes were cloned from *Drosophila* before the biochemical role of the encoded proteins was elucidated.

In *Drosophila* the paradigmatic homeotic genes are *Antennapedia*, mutations in which cause the transformation of an antenna to an extra leg, and *bithorax*, mutations in which cause the transformation of the halteres (balancing organs) into a second pair of wings.

When these homeotic genes were sequenced, it became clear that at or near the 3′-end of the gene, each contained a strongly conserved region, which has been termed the "homeobox". The homeobox is a 183-base-pair DNA sequence encoding the homeodomain. The protein products of these homeobox genes are transcription factors and the homeodomain forms a 61-amino-acid DNA-binding domain. Genes containing this sequence are commonly referred to as homeobox genes and the proteins they encode as homeodomain proteins. Sequence identity between mammalian and insect homeoboxes is typically 65–80% at the nucleotide level. Homeoboxes exist at least as far back in evolution as the Hydrozoans and they have been found in plants too.

In humans and mice, approximately 40 homeobox genes are arranged in four clusters on separate chromosomes. The organisation and spacing of the clusters is very similar; presumably the four clusters arose from duplication of the archetypal cluster. The genes in these clusters are commonly referred to as *Hox* genes. However, it should not be forgotten that there are also many non-*Hox* homeobox genes outside these clusters.

The homeobox genes have been extensively studied with regard to their role in the developing nervous system. In addition, it has been reported that various homeobox genes are involved in limb development and hematopoiesis. It is also clear that many homeobox genes are expressed in the skin.

2.1 Mechanism of action of homeodomain proteins

Being transcription factors, homeobox proteins operate by altering the level of transcription of other genes. In order to understand the biological effects of these proteins, it is necessary to determine the nature of the "target" genes in a given tissue. Additionally, as many homeobox genes are expressed in more than one location in the adult and during development, it seems unlikely that they are the only factor controlling tissue-specific expression. Interactions with other transcription factors and regulatory mechanisms seem to be required to control expression of different genes in different tissues.

The homeobox is a DNA sequence that encodes a conserved DNA-binding domain, the homeodomain.

Several of the amino acid residues present in the amino-terminal arm that are expected to give homeodomain proteins specificity are not predicted to contact DNA in monomeric binding, suggesting that they contribute to specificity by interacting with other proteins. It is also possible that the formation of protein heterodimers or higher order protein structures will alter the DNA-binding conformation, thereby allowing these residues to directly contact DNA.

2.2 Homeodomain protein target genes

The homeobox genes encode transcription factors; transcription factors act by altering the expression of their target genes.

A large gap in homeobox research has slowly begun to be filled in recent years, as finally a number of the target genes for homeodomain proteins have been characterised. To date, only a few primary structural mammalian target genes have been shown to be modulated by homeobox proteins. These include (1) the cell-surface morphoregulatory molecule NCAM (neural cell adhesion molecule), (2) the extracellular matrix morphoregulatory molecule cytotactin (also known as tenascin and hexabrachion), (3) the γ-globin gene, and (4) the thyroid transcription factor-1 gene.

3. THE HOMEOBOX GENES AND THEIR POSSIBLE ROLE IN DISEASE

A number of reports regarding possible connections between homeobox genes and diseases are summarised in Figure 3.1. What still remains to be uncovered in these diseases are the target genes, which the defective or missing homeodomain protein misactivates, or fails to activate (at the correct level or at all).

3.1 Homeobox genes in skin and skin appendages

Many homeobox genes are expressed in skin; moreover, it has been proposed that the homeobox genes in skin may provide the basis of a

Homeobox gene implicated	Disease or indication
Pax-5	The t(9;14) (p13;q32) chromosomal translocation associated with lymphoplasmacytoid lymphoma involves the Pax-5 gene.
Pax-6	Mutations in Pax-6 cause aniridia, Peter's anomaly, and autosomal dominant keratitis (ADK) in man, the *small eye* phenotype in mice, and the eyeless phenotype in *Drosophila*.
HUP2/Pax-3	Mutations in *pax-3* cause Waardenburg's syndrome in man, *splotch* phenotype in mice. However, Cases of Waardenburg's syndrome *not* linked to *Pax-3* have also been reported. *Pax-3* is also rearranged in the t(2;13) (q35;q14) translocation in pediatric solid tumour alveolar rhabdomyosarcoma.
Pax-1	A mutation in *Pax-1* has been reported in a patient suffering from spina bifida.
EMX2	Sporadic mutations were seen in patients affected with schizencephaly.
Ncx/Hox11L.1	Hox11L.1-deficient mice develop megacolon.
Hox-A4	Transgenic mice overexpressing Hox-a4 develop megacolon.
GSCL	The DiGeorge syndrome minimal critical region at 22q11.2 contains seven genes, including a goosecoid-like (GSCL) homeobox gene.
Hox-b2	Hox-b2 knockout mice display a phenotype resembling Bell's Palsy and Moebius Syndrome in man.
Msx1	Mutation in Msx1 leads to an autosomal dominant form of tooth agenesis.
Msx2	Mutation in Msx2 leads to an autosomal dominant craniosynostosis.
Hox-D13	Synpolydactyly in heterozygous individuals.
Dss1, DLXS, or DLX6 (all mapped to 7q21.2-q22.1)	Split hand-foot malformation, a form of ectrodactyly.
hLH-2	The human LIM-Hox gene hLH-2 is aberrantly expressed in chronic myelogenous leukaemia.
Hox-A9	Fusion of the nucleoporin gene to Hox-A9 by the t(7;11) (p1S;p1S) translocation causes human myeloid leukaemia.
Msx1	Msx1-deficient mice show cleft palate and abnormalities of craniofacial and tooth development.

Figure 3.1 Homeobox genes implicated in mammalian diseases or conditions.

Many homeobox genes are expressed in mammalian skin.

code, specifying the location of skin appendages (feathers, scales, whiskers, hair, and glands). Interestingly, in connection with this, two molecules intimately associated with skin and skin appendage development, tenascin and neural cell adhesion molecule (NCAM), are believed to be regulated by homeodomain proteins. Several non-*Hox* homeobox genes have been characterised in human foetal and adult skin, where they are differentially expressed. They were isolated by screening a foetal skin cDNA library for homeobox sequences.

3.2 Homeobox genes in epithelial-mesenchymal interactions

In skin, development of the various cutaneous structures (for example, hair, feathers, and sweat glands) is controlled by interactions between epithelial sheets and the adjacent mesenchymal cells. Excellent examples come from experiments in which the epithelium and underlying mesenchymal tissue are separated and then recombined in a different manner. From such studies it was apparent that the mesenchyme controlled the specificity of induction, because the same ectoderm developed differently depending on the source of the underlying mesoderm.

Strikingly, the signals and mechanisms involved are strongly conserved, even across species. Thus, when chicken corneal ectoderm is placed over chicken skin mesoderm, feather buds develop, just as they do if the chicken corneal ectoderm is placed over mouse skin mesoderm. Thus, chicken ectoderm is able to respond to the mesenchymal signal, which is seemingly conserved between chicken and mouse.

A homeodomain protein is involved in the specification of ectodermal structures in the skin.

Extending these results into the molecular arena, a 470-base-pair regulatory sequence from the *Xenopus Xdll-2* homeobox gene confers appropriate expression on a *lac-Z* reporter gene in the ectodermal component of structures derived from epithelial–mesenchymal interactions in transgenic mice. As was observed with chicken–mouse combinations, where chicken ectoderm responds to murine mesenchymal signals, the *Xenopus* sequence directed expression in structures not present in *Xenopus*, including the hair follicle and mammary gland, again suggesting that signals are conserved across species. Other studies have implicated the related homeobox gene *DLX4* in regulating epithelial–mesenchymal interactions in human placenta.

3.3 Patterning in skin diseases

Many skin diseases are patterned along Blaschko's lines.

In 1901, Alfred Blaschko published the results of a study of more than 140 patients, suffering from various skin diseases. Blaschko meticulously documented the patterns of the lesions on these patients; the diagrams he drew of these lesions (see Figure 3.2) revealed a series of lines, hence Blaschko's lines. The lines do not correspond to blood vessels, nerves, lymphatic drainage, or any other known structure.

The dermatological lesions of many diseases follow Blaschko's lines, while others do not. Other skin diseases in which lesions occur symmetrically about the midline are also known.

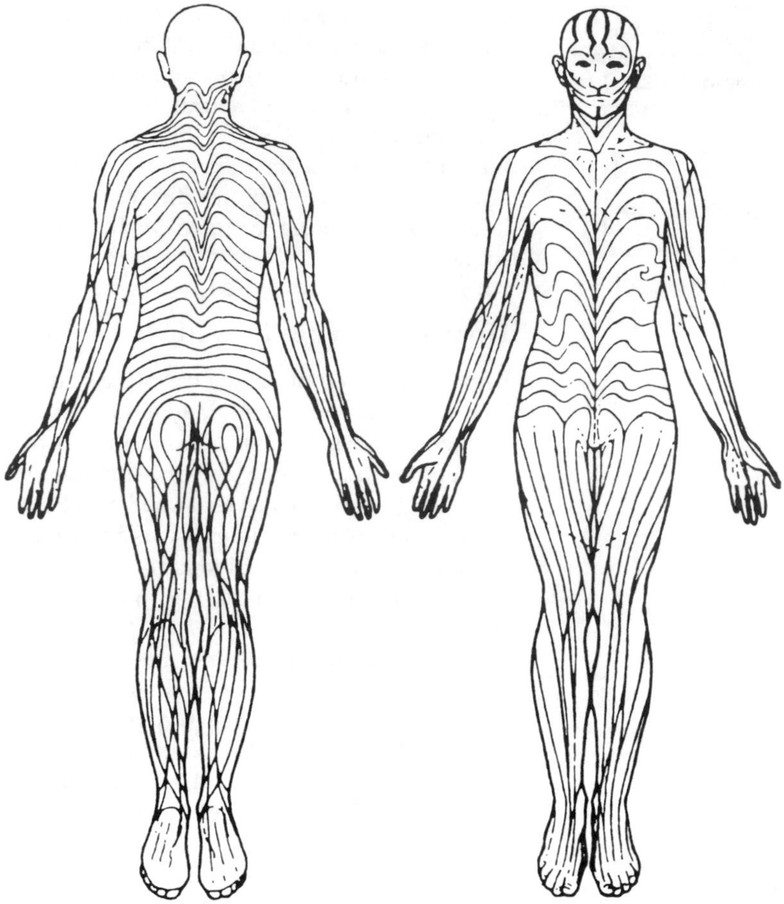

Figure 3.2 Blaschko's lines.

3.4 Homeobox genes in patterning

Homeobox proteins are known to be involved in forming the body axis of higher animals. In perhaps the most well-known embryological experiment, Spemann demonstrated that if the dorsal lip of the developing blastopore is transplanted to the opposite side of the embryo, a new, complete, second body axis resulted. One of the proteins central to this induction has now been characterised and it is the homeodomain protein "goosecoid". Goosecoid itself can largely reproduce the Spemann effect; if goosecoid mRNA is injected into the ventral side of an embryo, where none is normally present, then a secondary body axis develops.

3.5 Homeobox genes in the patterning of skin diseases?

It is perhaps not a giant leap to propose that homeobox genes might be involved in or provide an explanation for Blaschko's lines and/or the other patterns in which skin lesions have been described. Dermatologists have long used patterning in diagnosis and exploring the possibility of a molecular link seems worthwhile.

4. Other examples of human diseases caused by transcription defects

Two well-known examples of human diseases caused by transcription factor defects are Burkitt's lymphoma and Philadelphia-positive leukaemia. In Burkitt's lymphoma, there is a chromosomal translocation of a transcription factor gene, whereas in Philadelphia-positive leukaemia there is inappropriate activation of a transcription factor because of the formation of an abnormal fusion gene (the BCR-ABL fusion gene). These and other examples will now be discussed.

4.1 Burkitt's lymphoma

In Burkitt's lymphoma, a chromosomal translocation alters a transcription factor gene.

Burkitt's lymphoma was first described in the late 1950s as a malignancy in African children; it is quite rare outside of Africa, though a related "American" form of the condition is now known. Several chromosomal translocations have been identified in victims of the cancer. These translocations cause molecular rearrangements of immunoglobulin genes and the *c-myc* oncogene.

The *c-myc* oncogene involved in the translocations encodes a transcription factor that regulates expression of other genes. Specifically, the encoded protein binds to the consensus DNA sequence 5′-CACGTG; target genes containing this sequence in their promoters include p53, ornithine decarboxylase, and α-prothymosin. In Burkitt's lymphoma, these translocations cause transcriptional activation, leading to abnormally high levels of c-myc protein and aberrant over-expression of its target genes, leading to the malignancy.

4.2 Philadelphia chromosome

The so-called "Philadelphia chromosome" was the first genetic change to be associated consistently with leukaemia. It has been heavily studied ever since and is now quite well understood at the molecular level. The Philadelphia chromosome results from the aberrant fusion of two genes, one on chromosome 9, the other on chromosome 22. The BCR-ABL fusion gene results and this fusion gene encodes an aberrant transcription factor. The leukaemia then results because the aberrant transcription factor aberrantly regulates its target genes.

4.3 NF-κB

One transcription factor that has been heavily studied is nuclear factor-κB (NF-κB). It is now clear that this protein is found in many tissue and cell types and is believed to be involved in the expression of cytokines, chemokines, growth factors, cell adhesion molecules, and class I MHC molecules (see chapter 4). Inappropriate activation of NF-κB is believed to lead to inflammation in several human conditions, including autoimmune diseases, arthritis, asthma, septic shock, and lung fibrosis. Inappropriate *inactivation* has been linked to apoptosis and delayed cell growth.

4.4 NKX2–5 and heart disease

The homeobox gene *NKX2–5*, which encodes a transcription factor, has been implicated in a type of congenital heart disease.

Another transcription factor linked to human disease is the homeobox protein NKX2–5, which has been linked to a congenital heart condition. The disease was mapped to human chromosome 5q35, a region containing the *NKX2–5* gene. Three mutations in the gene have been identified; all three appear to alter the homeobox protein's ability to bind to DNA. Two impair the binding and one is thought to strengthen it. Presumably then, the altered proteins cause aberrant levels of transcription from their target genes, leading to the congenital heart anomalies.

4.5 *PAX* (paired box) gene defects

The *PAX* (paired box) genes encode a family of transcription factors related to the homeobox genes. *PAX* gene products have roles in development within several cells and tissue types, including cells of neural, myogenic, and lymphoid tissue.

Defects in several *PAX* genes have been linked with various tumours. Specifically, chromosomal translocations involving *PAX* genes have been linked with the myogenic soft tissue cancer alveolar rhabdomyosarcoma. As a result of the translocations, in-frame fusions are made from the *PAX* gene with other genes to make aberrant fusion proteins that bind aberrantly to target gene promoters and cause aberrant over- or under-expression of the proteins encoded by those target genes. Similarly, another translocation has been associated with the B-cell tumour lymphoplasmacytoid lymphoma. In this translocation, the intact *PAX5* gene is fused near an enhancer sequence, leading to deregulated expression of PAX5, and subsequent aberrant expression of its target genes.

Summary

Transcription factors have now been implicated in many human diseases. In addition, the homeobox genes were first characterised as a result of their ability to cause gross homeotic transformations in *Drosophila*. Neither should be surprising; transcription factors sit at the top of a cascade of their target genes. Defects in a transcription factor are likely to have wide-ranging effects; at the very least, all of the target genes will be aberrantly over- or under-expressed. Frequently, transcription factors regulate other transcription factors, magnifying these effects.

FURTHER READING

1. Latchman, D. S. (1997) How can we use our growing understanding of gene transcription to discover effective new medicines? *Curr. Opin. Biotechnol.* **8**, 713–717.
2. Peltz, G. (1997) Transcription factors in immune-mediated disease. *Curr. Opin. Biotechnol.* **8**, 467–473.
3. Barnes, P. J. and Karin, M. (1997) Nuclear factor κB – a pivotal transcription factor in chronic inflammatory diseases. *New Engl. J. Med.* **336**, 1066–1071.
4. Mundlos, S. and Olsen, B. R. (1997) Heritable diseases of the skeleton. Part I: Molecular insights into skeletal development – transcription factors and signalling pathways. *FASEB J.* **11**, 125–132.
5. Engelkamp, D. and Van Heyningen, V. (1996) Transcription factors in disease. *Curr. Opin. Genet. Dev.* **6**, 334–342.
6. Lemaire, P. and Kodjabachian, L. (1996) The vertebrate organiser: structure and molecules. *Trends Genet.* **12**, 525–531.
7. Lombardo, A. (1996) The organiser formation: two molecules are better than one. *BioEssays* **18**, 267–270.
8. Sorensen, P. H. B. and Triche, T. J. (1996) Gene fusions encoding chimeric transcription factors in solid tumours. *Semin. Cancer Biol.* **7**, 3–14.
9. Barr, F. G. (1997) Fusions involving paired box and fork head family transcription factors in the pediatric cancer alveolar rhabdomyosarcoma. *Curr. Top. Microbiol. Immunol.* **220**, 113–129.
10. Treisman, J., Harris, E. and Desplan, C. (1989) The paired box encodes a second DNA-binding domain in the paired homeodomain protein. *Genes Dev.* **5**, 594–604.

END OF UNIT QUESTIONS

1. Explain the significance of a defect in a transcription factor, compared to–for example–a muscle structural protein.
2. What is a homeotic transformation? List two that have been characterised in *Drosophila.*
3. How might a chromosomal translocation lead to aberrant expression of a transcription factor?
4. What is the homeobox?

Problems

1. Describe the ways in which NF-κB has been implicated in human diseases (see reference 3).
2. Conduct a search of the US National Library of Medicine's MedLine database, which is available on the Internet, for publications in the past two years implicating transcription factors in human diseases.

4 The Components of Transcription

OBJECTIVES

a) to introduce some of the key components of transcription

1. Overview

This chapter will introduce all the basic components of transcription *except* RNA polymerase, which will be discussed in the next two chapters, and histones, nucleosomes, and chromatin, which are discussed in chapter 7.

A promoter must be of at least a certain minimum size to ensure that it does not occur at random in the genome. This minimum size can be calculated, based on the size of the genome.

Additionally, if each gene needed a unique protein to bind to its unique promoter to switch it on or off, then at least half of the genome would need to be transcription factors. A better answer would be to use multiple transcription factors in some sort of combinatorial code (like telephone area codes). That way, many fewer transcription factors would be needed and multiple target genes could be activated by a given transcription factor.

Turning to the transcription factors themselves, how do they switch genes on and off? The basic DNA–protein interaction uses lock-and-key molecular complementarity, for want of a better description. That is essentially what the NMR and x-ray crystal structures of proteins bound to DNA show us: protein structures fitting into the minor and major grooves of the DNA double helix.

Even with transcription factors working in combination, there are still hundreds of them to control the thousands of genes. As we have seen already, they belong in large families of related proteins and the family members share a DNA-binding domain or structure. How then do the proteins recognise and distinguish the many DNA sequences seen in promoters?

2. Promoters

All genes within the genome contain at least one promoter. The promoter is a region of the gene upstream of the transcription start point on which protein complexes assemble to produce an RNA transcript. The initiation of transcription involves the specific assembly of large multi-protein complexes at the promoter.

With four different bases, the probability of a sequence of n bases occurring by chance is 1 in 4^n. From this and a knowledge of the size of the genome, the minimum size of a promoter can be established.

As a simple matter of statistics, a bacterial promoter must comprise at least 12 bases of DNA (not necessarily 12 *consecutive* bases) or it would likely occur by chance in the genome. Spacing between two or more elements can add further specificity. Twelve-base promoters are necessary because the probability of an 11-base promoter occurring by chance is 1 in 4^{11} = 1 in 4,194,304; thus, an 11-base sequence would be expected to occur probably once in the *E. coli* genome of 4.6×10^6 bases *by chance* alone. On the other hand, 4^{12} = 16,777,216 and a 12-base promoter is, therefore, unlikely to occur by chance in a genome of that size.

By the same argument, a human gene promoter would have to be at least 17 bases in length. There is a reasonable chance that a 16-base sequence would occur at random in the genome; $4^{16} = 4.29 \times 10^9$ and the human genome is approximately 3×10^9 bases. Extending the promoter by one base, $4^{17} = 1.72 \times 10^{10}$; thus, a 17-base-pair sequence is unlikely to occur at random.

To prevent promoter sequences occurring by chance, a promoter in *E. coli* must be a minimum of 12 bases in length while a human promoter must comprise at least 17 bases.

Of course, these calculations are simplistic and assume a *random* distribution of nucleotides in the genome, which is *not* true. However, they do provide a reasonable approximation and a foundation.

Next, consider also the implications of each gene having its own unique 12-base promoter sequence in *E. coli* or 17-base promoter sequence in man. If every human gene had a 17-bp promoter sequence and unique protein that bound it to switch it on and off, at least half the genome would need to be transcription factors to control the other half of the genome. Clearly, that is not going to make a very efficient genome or transcriptional system.

While 17 bases are necessary to ensure specificity, the 17 bases do not need to be *consecutive*. Three 6-base sequences would guarantee sufficient specificity. Furthermore, the spacing and order of the 6-base elements could provide further degrees of specificity (Figure 4.1). Additionally, by using elements and transcription factors that bind them in certain *combinations*, (1) rather fewer transcription factors are necessary and (2) *groups* of genes could be switched on or off in such a combinatorial system by altering the activity of a *single* transcription factor.

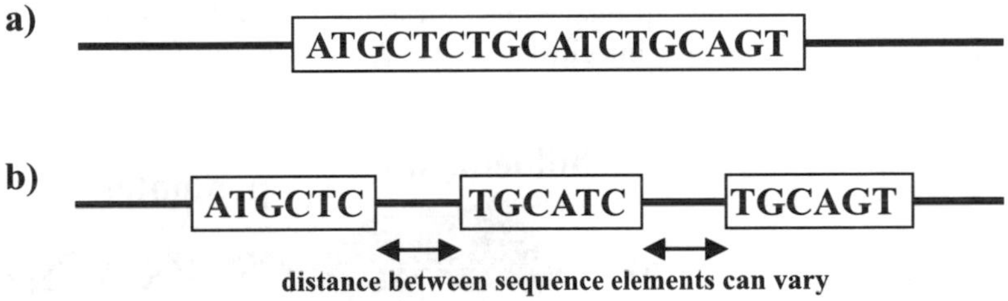

Figure 4.1 Alternative promoter designs. (a) An 18 bp promoter versus (b) a promoter of three 6 bp sequences. The combinatorial promoter allows fewer transcription factors to control genes and allows for coordinated control.

2.1 Bacterial promoters

As will be discussed in chapter 5, all transcription within bacteria is performed by a single type of RNA polymerase. The purified enzyme, or core enzyme, consists of four subunits, two copies of the α subunit and one each of a β and β' subunit. Together these subunits bind with low affinity to any region of double-stranded DNA. It is thought that by binding loosely, releasing momentarily, and then binding again, an RNA polymerase can efficiently explore or scan the chromosomal DNA. For specific recognition of bacterial gene promoters, the RNA polymerase core enzyme will bind with one of a set of accessory proteins referred to as *sigma* (σ) factors.

When this complex binds a specific promoter, the bound *holoenzyme* (RNA polymerase plus σ factor) causes the double-stranded DNA to unwind or "melt", so that transcription can begin. The σ factor is only needed at the start of transcription and is released after the first few nucleotides of the RNA have been added (Figure 4.2).

Bacterial promoters typically contain a Pribnow box, 5′-TATAAT, at –5 and the sequence 5′-TTGAC at around –35.

The most extensively studied bacterial promoters are those of *Escherichia coli (E. coli)*. Bacterial promoters typically span about 40 bases and include a common sequence of six bases upstream (5′) of the transcription initiation point. This sequence has become known as the Pribnow box after the researcher who first described the sequence. The base in the DNA that corresponds to the first nucleotide of the RNA transcript is designated +1; the bases upstream of this (5′) are described using negative numbers, there being no "zero" position. The Pribnow box is typically located five to eight bases from the initiation site (+1). The consensus sequence of the Pribnow box is 5′-TATAAT, though this varies somewhat; however, the first A and the last T are nearly always present. Within bacterial promoters there is a second consensus sequence, 5′-TTGAC, which is typically found around the –35 position in the promoter. This sequence appears to be important for the rapid and accurate initiation of transcription of most bacterial genes.

As described earlier, recognition of a promoter by the bacterial RNA polymerase and binding of this enzyme to it are mediated by a σ factor. A large number of factors have now been identified and sequenced. Each σ factor permits RNA polymerase to bind to promoters that contain different consensus sequences in the –35 region.

2.2 Eukaryotic promoters

Eukaryotic genes are often very different from those seen in prokaryotes. First, as a matter of definition, eukaryotes can be unicellular (*e.g.*, yeast) or multi-cellular (*e.g.*, humans). The key

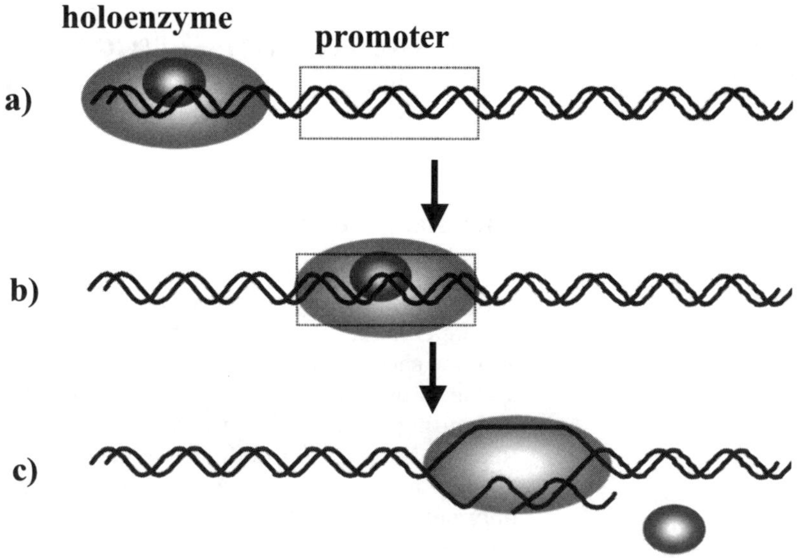

Figure 4.2 Initiation of transcription.
a) Holoenzyme binds to any DNA with low affinity.
b) When the enzyme recognises a promoter, binding affinity increases.
c) At the promoter the DNA will "melt" (*i.e.*, the strands will be separated) allowing transcription to begin. Immediately after initiation, the sigma (σ) factor leaves the transcription complex.

characteristic is that the cells have a true nucleus, a distinct organelle with a nuclear membrane containing the chromosomes.

In eukaryotes, promoter activation is far more complicated and complex because of the increased number and type of promoters that are present in the larger genome. Additionally, the eukaryotic cell needs to alter gene expression during development and in a tissue-specific and cell type-specific manner.

The problem of this complexity cannot be solved entirely by the binding of specific proteins directly to the promoter DNA, as specificity determinants at the promoter are limited and cannot provide a sufficient number of permutations. One solution to this problem appears to be the binding of protein factors to regions away from the immediate vicinity of the promoter (usually upstream of the promoter). Such *cis* interactions permit sequences that are situated a significant distance from the transcription start site to participate and influence the control of the initiation of transcription. Proteins bound to such sequence-specific sites can be brought into close physical proximity, and assemble into specific activation complexes at the promoter as a result of DNA looping.

These *cis* interactions are of two types. One involves the binding of proteins to DNA sequences called "upstream elements", which are located 100–200 bp from the promoter, as defined by the TATA box. Related interactions also occur in prokaryotes. The second involves binding of protein factors to "enhancer" sequences (discussed below), which are found up to several thousand base pairs from the promoter, and tend to be confined to eukaryotes.

In contrast to prokaryotic promoters, eukaryotic promoters contain a variety of regulatory elements at varying distances from the transcriptional start point; presumably the nature, number, and spacing of the elements add further levels of specificity and selectivity. Additionally, as will be discussed in chapter 6, in eukaryotes, there are three kinds of RNA polymerase; mRNA is transcribed only by RNA polymerase II.

Eukaryotic RNA polymerase II promoters typically contain a TATA box (or Hogness box) at -25. Alterations in the TATA box lead to major decreases in transcriptional initiation.

Promoters for RNA polymerase II, like those for bacterial polymerases, are typically located on the 5′ side of the transcriptional start site. As in prokaryotic promoters, there are several important and conserved regions. The region positioned closest to the start point is referred to as the TATA box (also known as the Hogness box) and is typically positioned at around the –25 position. The consensus sequence is a six-nucleotide stretch comprising only A and T residues. The TATA box is present in nearly all eukaryotic genes from which mRNA is transcribed. Alterations in this TATA box sequence markedly affect the initiation of transcription, demonstrating the sequence's importance.

While the TATA box is nearly always necessary for efficient gene activation, it is often not the only element present in a promoter. Additional elements are commonly located between –40 and –110. Many promoters contain a so-called "CCAAT" box and some contain a "GGGCG" box.

As mentioned, some eukaryotic promoters are known that do *not* contain a TATA box, but many of these genes are not transcribed

While the TATA box is necessary, it is insufficient to ensure efficient gene expression.

at high rates. Most of these genes, however, contain a GC-rich stretch of 20–50 nucleotides located within the first 100–200 bases upstream (5′) of the transcription start site. The transcription factor, Sp1, recognises such GC-rich sequences, and genes lacking the TATA box may rely on these GC-rich sites and proteins bound to them to initiate transcription.

Systematic site-directed mutagenesis experiments have shown that each eukaryotic gene has a combination of regulatory *cis* elements that are uniquely arranged as to number, type, and spacing. Usually these elements are located within a few hundred base pairs of the transcription initiation site. These *cis* elements are binding sites for transcription factors that activate or repress transcription. Additionally, some genes contain enhancer sequences; these are also believed to be protein binding sites, but operate in a largely distance- and orientation-independent manner (discussed below).

Many other less common elements are found in promoters, including those that confer sensitivity to signals in the heat shock response, hormones, and growth factors. Overlapping or super-imposed binding sites for several transcription factors are seen in several promoters. In some genes, synergistic effects dependent on the strict spacing of adjacent *cis* elements have been observed. To add a further layer of complexity, some genes have more than one transcriptional start site, each being regulated by its own *cis* elements and transcription factors. Clearly the number of possible combinations between these mechanisms is large and is presumably necessary to cope with the complexities of eukaryotic cell life, tissue and cell specialisation, and the size of the genome.

2.2.1 *MHC class I gene promoters*

One of the central dogmas of modern immunology is that all nucleated cells normally express major histocompatibility complex (MHC) class I molecules on their surface. For most purposes, this means all cells except erythrocytes (red blood cells), though there are actually some other cell types that are deficient in or completely lack class I proteins on their surface, including cells in the central nervous system (CNS), neuroblastomas, sperm, some cells of the trophoblast, and choriocarcinoma cell lines derived therefrom.

Antigenic peptides bound to class I molecules form a complex that is recognised by the T-cell receptors (TCRs) on the outer surface of T-cells. The six peptide-binding pockets (A–F) of the class I molecules vary in shape, depth, and chemical environment in the different allelic proteins. This sequence variation influences the affinity for distinct peptides and determines the peptide repertoire that each allelic MHC protein can present.

T-cells "inspect" peptides presented in MHC molecules and kill cells that present foreign or defective peptides (Figure 4.3). However, an infected or cancerous cell that does not express MHC molecules would escape detection by T-cells and would not be killed by them. Loss of class I expression will therefore paralyse host defences against infectious agents or tumour cells. Some viruses and cancer cells do exactly

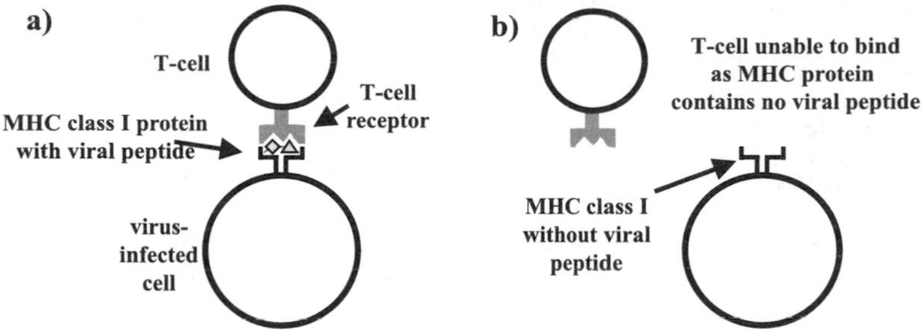

Figure 4.3 MHC class I protein molecule presenting a viral peptide to a T-cell.

a) virally infected cell presenting a peptide to a T-cell.
b) in a non-infected cell, the MHC protein is "empty" and the T-cell will not interact.

this; that is, they switch off expression of MHC molecules, to make the cells "invisible" to T-cells.

Studies on class I-negative tumour cells have revealed mutations in the enhancer A (Enh A) region, illustrating that control is at the level of transcription. Down-regulation or complete, sometimes selective, loss of class I antigen expression has been reported in various tumour cells; this down-regulation can be locus- or allele-specific. In some cases, this down-regulation or loss is associated with altered levels of the transcription factors that bind these regulatory elements.

Decreased expression of HLA class I genes has also been found in a high percentage of melanoma cells and neuroblastoma cells: HLA-A, -B, and -C can all be down-modulated by N-myc in neuroblastoma cells, which is believed to be mediated by inactivation of Enh A. This suggests that the selective down-regulation of class I antigens or the absence of these molecules in certain tumour cells is linked to the transcriptional regulation of the genes.

Infected or cancerous cells do not escape scot-free, however. To complement the action of T-cells, natural killer (NK) cells kill cells that have no MHC molecules on their surface, because such cells must also be abnormal.

The promoters of the MHC class I genes have been extensively studied and contain examples of many of the elements and features discussed in the preceding paragraphs. The MHC is a region of approximately 4 megabases (4×10^6 bases) on the short arm of human chromosome 6. Major histocompatibility complex (MHC) proteins, also referred to as HLA (human leukocyte antigen) molecules, are proteins present on the outer surface of antigen-presenting cells (APCs). There are two major types of MHC molecules, class I and class II, and they present antigens to the two primary types of T-cells, cytolytic T-cells, and T helper cells, respectively.

The HLA class I genes are the most polymorphic genetic system yet characterised; that is, across a population, there are many minor sequence variations in the proteins. To date, several hundred different HLA-A, HLA-B, or HLA-C alleles have been cloned and sequenced. The mature molecule comprises a glycosylated 43 kDa heavy chain, which is an integral membrane protein, non-covalently associated with the non-polymorphic β_2-macroglobulin (β_2M) molecule.

In DNA terms, "*cis*" means located on the same, physically contiguous DNA molecule. "*Trans*" means located on another DNA molecule.

The genes encoding the HLA-A, -B, and -C class I molecules are constitutively transcribed in most adult cells. Modified expression levels of these genes have frequently been observed in various types of human and animal tumour cells. The class I proteins are found on the cell surface of almost all nucleated cells, though their expression is regulated both developmentally and in a tissue-specific manner. Expression is modulated by a variety of immunologic stimuli, including cytokines, viral infection, and tumour cell transformation. At least one mechanism of this regulation and modulation is at the level of transcription.

Transcriptional control of the MHC class I genes is mediated largely through the class I regulatory sequence (CRC) and the many transcription factors that bind it.

Transcriptional regulation of class I genes is controlled via a series of *cis*-regulatory elements located in the 5′ flanking upstream region and the various transcriptional factors that bind to them. The promoter region of MHC class I genes has been studied in great detail and is known to contain two conserved *cis*-acting elements, referred to as enhancer A (Enh A) and enhancer B (Enh B). The enhancer A sequence, also referred to as the class I regulatory complex (CRC), is located between –200 and –150 bases from the transcriptional start site and plays a major role in transcriptional regulation.

The CRC contains a series of *cis*-acting elements, including the κB sites, κB_1 and κB_2. The κB_1 motif is a highly conserved palindromic sequence (5′-GGGGATTCCCC), located upstream of the IFN-responsive element (IRSE), and is bound by many transcription factors, including NF-κB. Additionally, the CRC contains an imperfect copy of the κB_1 sequence a few bases upstream, the κB_2 sequence (Figure 4.4). In a variety of cell types these elements appear to be important controllers of class I gene expression.

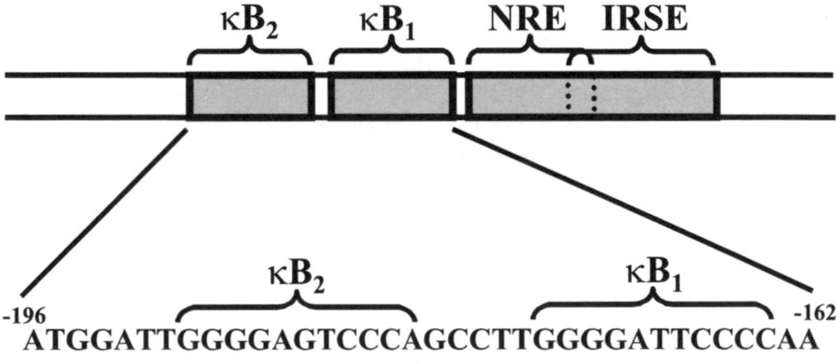

Figure 4.4 The major histocompatibility complex (MHC) class I regulatory complex (CRC) contains several closely adjacent and overlapping DNA sequences to which transcription factors bind, including the heavily studied κB sites, to which the transcription factor NF-κB binds.

2.2.2 *NF-κB*

NF-κB is an inducible DNA-binding factor that was initially characterised as a positive transactivator of immunoglobulin (Ig) κ gene. NF-κB has since been shown to be a key regulator of various genes encoding proteins of the immune system, including the MHC class I genes. The κB-binding protein family comprises at least five subunit proteins, p50, p65 (RelA), p52 (p50B), c-Rel, and RelB, which can form homo- or heterodimers between themselves. The various dimeric complexes differ in their binding capability for certain κB sites, transactivation potentials, kinetics of nuclear translocation, and levels of expression in tissues.

Members of the κB-binding protein family share the Rel homology domain (RHD). It has been shown that the expression and activation of the different complexes are differentially regulated through cell-specific and temporally distinct pathways. Crystal structures of the p50 homodimer bound to DNA have been determined and DNA binding is apparently mediated via the RHD, which folds into two domains, similar to that seen in the immunoglobulin superfamily. It seems likely that the other members of the Rel family, sharing the RHD, will bind DNA in a similar fashion, protein sequence differences in the members being responsible for their varying DNA-binding specificity.

3. ENHANCERS

Another element found in eukaryotic DNA that controls transcription is a sequence known as an enhancer. Enhancers were originally characterised as having the ability to stimulate transcription practically regardless of their orientation or distance relative to the transcriptional start site. These were clearly unexpected properties based on the previously characterised bacterial promoters, where spacing and orientation were crucial.

Enhancers are DNA sequences that stimulate transcription, almost regardless of orientation and distance from the transcriptional start site.

Enhancers are typically found more than 100 bases upstream of (5′ to) the transcriptional start site. However, enhancers have also been found thousands of base pairs upstream or downstream, and within introns (as in the immunoglobulin κ gene). As examples, the wing margin enhancer of the *Drosophila* cut locus is found 85 kb upstream of its promoter, the murine immunoglobulin Hμ core enhancer is located within the second intron of the gene, and T-cell receptor α-chain gene enhancer is 69 kb downstream of the promoter. Similarly, expression of the mouse POU domain protein Tst-1/Oct6 is controlled in part by an enhancer located in the 5′ flanking region, 5 kb upstream of the transcriptional start site.

It is generally accepted that these DNA elements, which are located relatively far from the transcription start site, become physically close to the start points by loops being formed in the DNA. Thus, DNA sequences some distance apart on the linear chain can become physically very close to each other, allowing protein–protein interactions.

It is currently believed that enhancers operate through several mechanisms. First, in common with the promoter elements discussed already, enhancers are *cis* elements that are bound by sequence-specific DNA-binding proteins. Second, co-activators can interact with the DNA-bound transcription factors (*i.e.*, without themselves binding DNA). Such protein–protein interactions can occur between DNA-bound and associated-but-not-DNA-bound proteins and components of the basal transcription machinery, that is, RNA polymerase and its associated proteins. DNA looping can bring the RNA polymerase and promoter–DNA complexes into close proximity. Finally, some degree of chromatin remodelling appears to occur and this also influences transcription.

Several transcription factors are known to covalently modify proteins, especially by phosphorylation or acetylation. Many transcriptional activators and co-activators possess histone acetyltransferase activity, whereas some transcriptional co-repressors have histone deacetylase activity. Thus, some transcription factors can potentially acetylate or deacetylate both histones and other transcription factors as part of the overall regulation of the transcription process.

Packaging of DNA and chromatin will be discussed in chapter 7.

3.1 Mechanism of action of enhancers

Several models have been proposed to explain how enhancers might operate to stimulate transcription from a gene. These include looping, scanning, and the so-called "facilitated tracking" model; these will now be briefly discussed.

Looping of DNA to bring the promoter and distant enhancer into close proximity or physical contact is perhaps the simplest model of enhancer operation. However, as the distance between the enhancer and promoter increases, even allowing for all the superstructure of chromatin, the chances of the sequences meeting and interacting fall off quickly.

A further proposed model is that of DNA scanning, in which enhancer-binding factors would bind to their DNA recognition sequences and then "scan" along the DNA until they reached their promoter. However, promoters and enhancers can be separated by one or more other active genes and such a mechanism would require "removal" from the DNA between the enhancer and promoter of other transcription factors and DNA-binding proteins. Additionally, such a model fails to address how an enhancer on one chromosome could activate transcription from an allelic promoter on another paired chromosome, as occurs in transvection. Transvection is a phenomenon in which the expression of a gene can be controlled by its homologous counterpart, presumably due to pairing of alleles in diploid interphase cells. It was first characterised in *Drosophila*, where it occurs in many genes.

The so-called "facilitated tracking" model incorporates elements from both the looping and scanning concepts. The model posits an enhancer-bound complex containing DNA-binding factors and co-activators that "tracks" along the DNA until it reaches its promoter, at which point a stable structure is formed.

3.2 Enhancer elements and DNA looping

In prokaryotes, a specificity factor (sigma (σ) factor, described above) binds to the core polymerase in solution and directs the resultant holopolymerase to the relevant class of promoter. The accessibility of a given promoter to the polymerase is modulated by protein activators and repressors that bind at or near the promoter to facilitate or inhibit the binding of the polymerase and the subsequent transcription initiation. The ability of these regulatory proteins to interact with a promoter is a direct result of their binding affinity for specific DNA sequences in the promoter region.

The DNA binding sites for regulatory proteins in prokaryotic promoters are of two basic types. Some are located close enough to the promoter sequence to permit direct protein–protein interaction to cause a regulatory effect. The second type of protein binding site may be located up to 200 base pairs from the promoter. These sites presumably interact with the transcription complex by forming a loop allowing protein–protein interactions. A notable example of this mechanism involves promoters that require the σ factor 54. At these promoters the nitrogen regulatory protein C (NTRC) activator protein binds near position –110 and cannot contact the polymerase without looping of the intervening DNA.

Regulatory proteins bound at enhancers and upstream elements are thought to participate in the formation of the active transcription initiation complex by looping of the DNA to permit interaction with the promoter.

The effectiveness of DNA looping in bringing protein factors to the promoter at high enough concentrations and with appropriate orientations is a quantitative problem and will depend on the flexibility and conformation of the intervening DNA. Both DNA flexibility and conformation are determined by structural features inherent within the DNA sequence as well as the effect of bound complexes, such as nucleosomes.

4. TRANSCRIPTION FACTORS

Transcription factors are DNA-binding proteins that mediate effects on gene expression.

Transcription factors are DNA-binding proteins that recognise *cis*-regulatory elements of target genes and are regulators of gene transcription. There are two classes of transcription factors regulating transcription in eukaryotes: general transcription factors - components of the basal transcriptional machinery, which will be discussed in the next chapter - and sequence- and tissue-specific transcription factors.

At a gross level, transcription factors can be regulated at two levels, concentration (amount) and activity. As a general matter, the concentration or amount of a transcription factor in a cell can be regulated at several levels: transcription, RNA processing, mRNA degradation, and translation.

During evolution, several ways of regulating the activity of transcription factors and proteins in general have developed, including phosphorylation, acetylation, nuclear localisation, DNA binding, and *trans*-activation. DNA-binding activity of transcription factors can be regulated by phosphorylation (*e.g.*, heat shock factor and serum response factor) or dephosphorylation (*e.g.*, PKC-dependent threonine dephosphorylation of c-Jun). Other mechanisms include O-glycosylation of – for example – the transcription factor Sp1. Sp1 is a ubiquitously expressed transcription factor that is particularly important in the regulation of TATA-less genes that encode housekeeping proteins.

Perhaps surprisingly for intracellular proteins, glycosylation is common in transcription factors that operate with RNA polymerase II. The exact functional role of the glycosylation of Sp1 remains to be determined.

Additionally, the activity of some transcription factors is altered by ligand binding, as is seen in the steroid hormone receptor superfamily. This will be further discussed below.

These and other mechanisms allow transcription factors to effect the controlled expression of specific subsets of the genome in different cells at different times in response to different signals. Additionally, when analysing this issue, it should be remembered that a protein can very much affect transcription without itself actually binding DNA. A protein could act as a bridge between DNA-binding transcription factors and the basal transcription machinery, for example. Many transcription factors occur in distinct families that share common structural features, especially the DNA-binding domain. Some examples will now be discussed.

4.1 Steroid hormone receptors

The steroid hormone receptors share considerable sequence homology and bind DNA via a zinc-finger-containing domain.

Steroid hormones effect changes in gene expression by passing through cell membranes, binding to cytoplasmic receptors, and causing their activation. These activated receptors then move from the cytoplasm to the nucleus and alter gene expression (Figure 4.5). This signal transduction pathway is unusual in that the hormone ligand

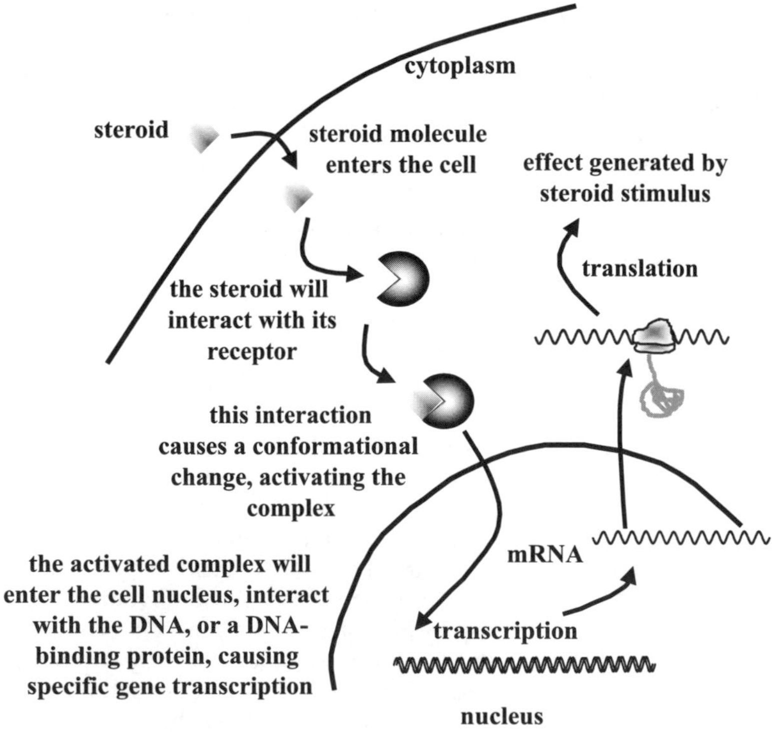

Figure 4.5 Steroid hormone receptor mechanism of action.

forms an integral part of the protein complex involved in DNA binding and transcriptional activation.

The steroid receptor superfamily of transcription factors includes not only the receptors for the steroids, but also those for thyroid hormone, retinoids, and vitamin D. Additionally it contains many receptors for which the ligand remains unknown; these receptors are generically referred to as "orphan receptors".

The various members of the superfamily share considerable sequence homology and three structural domains or motifs. Specifically, all the members share (1) an amino-terminal region, believed to be involved in modulating gene expression, (2) a zinc finger DNA-binding domain, and (3) a less conserved carboxyl-terminal region, believed to be involved in ligand binding and receptor dimerisation. Binding of the hormone to the carboxyl-terminal region induces a conformational change in the receptor and allows it to bind DNA.

The DNA sequences to which the receptor binds are referred to as hormone response elements and are 12–18 bases in length. The elements are partially palindromic ("mirror" images by base pairing), there being two "half-sites" separated by a variable spacer sequence. Both the sequence of the response element and the orientation and spacing between the two half-sites are important in determining the specificity of binding. The binding of nuclear hormone receptors to their hormone response elements occurs as dimers; one receptor molecule binds to each half-site.

4.2 Jun: transcription factor and oncoprotein

The protein v-jun was first identified as the transforming protein of the sarcoma virus 17. Since then, three mammalian homologues have been characterised. The three jun genes encode proteins that bind to DNA and positively or negatively regulate transcription of genes. As a result of transcription factor "cross-talk", jun also can affect transcription from genes regulated by other types of transcription factors, including steroid hormone receptors.

Activator protein-1 (AP-1) is an inducible transcription factor that comprises several protein complexes that include the proteins encoded by the *fos* and *jun* gene families. Many cellular and viral genes contain AP-1 binding sites in their promoters.

4.3 C/EBP and the "leucine zipper" structure

C/EBP is expressed in adipocytes, hepatocytes, and monocytes/macrophages. Target genes of the proteins are believed to include the acute phase response genes in hepatocytes and cytokine genes in monocytes/macrophages. C/EBP is now known to be a member of a family of transcription factors. Members of the family share three structural components: a carboxyl-terminal leucine zipper region, a basic DNA-binding region, and an amino-terminal *trans*-activating region.

The so-called "leucine zipper" motif is a structure involved in protein dimerisation. It is found in some transcription factors.

The leucine zipper motif is a protein-protein binding motif, which is present in a number of transcription-related proteins, including the yeast transcription activator GCN4, the oncoproteins jun, fos, and myc, and the C/EBP enhancer binding protein, as discussed in chapter 2.

Leucine zipper-containing proteins can form both homo- and heterodimers; the homo- or heterodimer then has two identical or different basic DNA-binding regions, respectively.

5. CHROMATIN

In addition to this interaction of receptors with proteins directly involved in transcription, there is evidence supporting a role in modulating chromatin structure. Nucleosomes and chromatin structure generally are discussed in chapter 7.

SUMMARY

Given four bases in DNA, promoters have to be of a certain minimum size or the sequence will occur by chance, given the size of the genome. Multiple elements are found in most promoters; such a scheme enables a code to be used, which reduces the number of required transcription factors and provides for co-ordinated regulation of genes through shared elements. Enhancers are DNA sequences that operate to stimulate transcription of genes *regardless* of their orientation and distance from the transcriptional start site; they are found upstream and downstream of genes and even within introns. Transcription factors are sequence-specific DNA-binding proteins that effect changes in gene expression. Selective gene expression is mediated largely through the actions of selective DNA-binding proteins, referred to generically as transcription factors. The known transcription factors readily fit into large families, sharing conserved DNA-binding and transcription-activating domains. The specificity of interactions must therefore be mediated by sequences outside these conserved regions.

FURTHER READING

1. Boulikas, T. (1994) A compilation and classification of DNA binding sites for protein transcription factors from vertebrates. *Crit. Rev. Eukar. Gene Expr.* **4**, 117–321.
2. Sarkar, N. (1997) Polyadenylation of mRNA in prokaryotes. *Annu. Rev. Biochem.* **66**, 173–197.
3. Pennypacker, K. R. (1995) Pharmacological regulation of transcription factor binding. *Pharmacology* **51**, 1–12.
4. Blackwood, E. M. and Kadonaga, J. T. (1998) Going the distance: a current view of enhancer action. *Science* **281**, 60–63.

5. Henderson, A. J. and Calame, K. L. (1995) Lessons in transcriptional regulation learned from studies on immunoglobulin genes. *Crit. Rev. Eukar. Gene Expr.* **5**, 255–280.
6. Udvardy, A. (1999) Dividing the empire: Boundary chromatin elements delimit the territory of enhancers. *EMBO J.* **18**, 1–8.
7. Beato, M. (1991) Transcriptional control by nuclear receptors. *FASEB J.* **5**, 2044–2051.
8. Boulikas, T. (1995) How enhancers work: juxtapositioning of DNA control elements by synergistic interaction of MARs. *Int. J. Oncol.* **6**, 1313–1318.
9. Novina, C. D. and Roy, A. L. (1996) Core promoters and transcriptional control. *Trends Genet.* **12**, 351–355.
10. Etienne, J., Brault, D. and Firmin, S. (1990) Cis-elements and DNA-binding proteins involved in transcription regulation. *Ann. Biol. Clin.* **48**, 681–694.

End of unit questions

1. Explain why the minimum size of a promoter in *E. coli* is 12 bases, given that the *E. coli* genome is 4.6×10^6 bases. What is the minimum size of a human promoter, given a genome size of 3×10^9 bases, and why?
2. If many transcription factors in a protein family share a conserved DNA-binding domain, how does each bind to different promoters?
3. How–at the level of transcription–might a cancerous cell "escape" detection by T-cells?

Problems

1. Discuss current ideas as to how enhancer sequences might function (see reference 4).
2. Discuss the role of the promoter and its component elements in the regulation of transcription (see references 9 and 10).

5 Bacterial Transcription

OBJECTIVES

a) introduce the bacterial RNA polymerase, its component parts and accessory factors, and its mechanism of action

1. Overview

Having described promoters, enhancers, and gene structure (chapters 2 and 4), we now come to the key enzyme that binds the promoter and transcribes RNA from the gene. RNA polymerase in *E. coli* has been extensively studied both for its own sake and as a model for the much larger and more complex eukaryotic enzyme, which will be discussed in the next chapter. As the next two chapters will demonstrate, the prokaryotic and eukaryotic enzymes have much in common.

2. RNA polymerase, generally

Enzymes that "copy" DNA by base pairing into DNA are known as DNA polymerases; those that "copy" DNA into RNA are RNA polymerases. RNA polymerases catalyse the nucleophilic attack of the 3′-hydroxyl group of the growing RNA chain on the α-phosphorus atom of the incoming ribonucleoside triphosphate molecule. No primer is needed for RNA synthesis, in contrast to DNA synthesis. An RNA polymerase can find an appropriate initiation site on double-stranded DNA, bind to it, separate the two strands in that region and begin generating a new RNA strand. The synthesis proceeds in the 5′→3′ direction according to the "instructions" given by the anti-parallel template strand. As the complex moves from the initiation site, the DNA strands re-anneal.

RNA polymerases produce mRNA by base pairing RNA bases with those in DNA.

The initiation of transcription by RNA polymerase requires the assembly of a multi-protein complex at the promoter region of the gene. Recently, using an adapted atomic force microscope, molecular-scale images have been obtained of the initial steps of transcription. The *E. coli* RNA polymerase's interaction with DNA has been studied. In the atomic force microscopy experiments, *E. coli* RNA polymerase was studied as it transcribed DNA templates. Transcription was observed by following the translocation of DNA through an anchored RNA polymerase protein complex. The incubation medium contained only three of the four ribonucleotides necessary to synthesize RNA. When the enzyme required the missing fourth base, it paused, enabling researchers to view the assembled complex. When the missing base was added to the medium, the DNA was observed being pulled through the anchored RNA polymerase. The transcription rates observed were approximately 0.5 to 2 bases per second under the conditions of the experiment. The produced RNA transcript was also identified.

3. Transcription in bacteria

Control of transcription is primarily at the level of initiation.

Control of gene expression is mediated in both prokaryotes and eukaryotes primarily at the level of the initiation of transcription. It is simply the most economical step to regulate.

Messenger RNA (mRNA) transcription from a given promoter is driven by a combination of two processes. First, the promoter must initially be bound by a polymerase. In this binding, the promoter sequence is in competition with other promoters and non-specific sites on the DNA. Second, the polymerase must move through the initiation phase of transcription into elongation as rapidly as possible to "clear" the promoter and make it available for reuse, that is, binding by another RNA polymerase molecule.

3.1 The bacterial RNA polymerase

The bacterium has only one RNA polymerase, comprising five polypeptides: two alpha (α) chains, one beta (β), one beta prime (β') and one omega (ω) chain. A sixth subunit, sigma (σ), is sometimes associated with the enzyme.

In prokaryotes, transcription is performed by a single type of RNA polymerase. The structure of this polymerase is strongly conserved across prokaryotes and has been extensively studied. The "core" polymerase comprises five polypeptides, two alpha (α) chains, one beta (β), one beta' (β') and one omega (ω) chain. A sixth subunit, sigma (σ), is also associated with the functional polymerase and its role is to direct the complex to initiate transcription at specific promoters. Together, these six subunits comprise the RNA polymerase holoenzyme. RNA polymerase also contains two zinc ions (Zn^{2+}) per complex, one associated with the β subunit and the other with the β' subunit. The total molecular weight of the enzyme complex is approximately 450 kDa. Additionally, the ω subunit has been implicated in the so-called "stringent response" (see chapter 10) and is involved in the co-ordination of RNA and protein synthesis.

The formation of active polymerase can be followed *in vitro* and complexes smaller than the complete holoenzyme have been identified. From these studies the following assembly order for the *E. coli* polymerase seems likely:

$$2\alpha \rightarrow \alpha_2 + \beta \rightarrow \alpha_2\beta + \beta' \rightarrow \alpha_2\beta\beta' \text{ (inactive)} \rightarrow \alpha_2\beta\beta' \text{ (active)}$$

In the final step, a conformational change occurs and the complex is then active. The functions of the subunits have been determined using mutants containing defects within the various RNA polymerase subunit genes. The β subunit contains the active site of the polymerase, where the nucleotides are joined to form RNA. It is also the β subunit that is affected by inhibitors of transcription such as rifampicin and streptolydigin. The β' subunit's role is DNA binding; unsurprisingly, the β' subunit is largely basic in nature, and therefore positively charged at physiological pH. The σ subunit is also involved in DNA binding; more specifically, its role is in specific promoter recognition. It only associates with the polymerase for part of the time and can be replaced by other σ factors.

The genome of *E. coli* comprises approximately 4,600,000 base pairs; within this genome there are approximately 2,000 genes and therefore gene promoters. The polymerase's task is to selectively seek out the gene(s) that need to be transcribed at a particular point in time.

3.2 How does RNA polymerase find promoters?

The initial stage of DNA–protein interaction could be just a random event, but based on mathematical analyses such random events could not account for the speed with which the polymerase molecules find and associate specifically with promoters. This conflict could be explained if, when the polymerase randomly binds to the DNA, it does not immediately dissociate again but remains "on" the DNA and tracks along the DNA until it recognises a promoter region. Through such theoretical calculations, it has been estimated that the polymerase can track along the DNA at up to 1,000 bases per second.

Recognition of a promoter by RNA polymerase is expected to occur within the DNA major groove, where the pattern of hydrogen bonds between the base pairs is identified. Other molecular interactions and steric effects are also involved in the recognition. The main sites described above are the –35 and –10 regions.

3.3 Initiation

At the point of initiation the DNA structure is altered from the "closed" to the "open" complex; the conformational change can be seen using DNase I footprinting, where there is a change in the nuclease sensitivity of the DNA.

When RNA polymerase locates a promoter, substantial conformational changes take place in both the polymerase and the DNA. These changes are referred to as the transition from the "closed" to the "open" complex. These conformational changes were first observed using DNase I footprinting; the DNA is differently susceptible to nuclease digestion, depending on whether it is open or closed.

For example, in the A3 promoter of bacteriophage T7, *E. coli* RNA polymerase protects the nucleotides between bases –56 to –5 in the closed complex. Thus, the enzyme covers about five complete helical turns of the DNA. On transition to the open complex, protection extends downstream to about position +20. In the closed complex, the DNA remains double stranded. However, this changes in the open complex and a section of the DNA becomes single stranded (melted), usually within the region between positions –9 and +3 (Figure 5.1).

In the closed complex there are usually contacts between the β, β', and σ subunits with both the coding and non-coding DNA strands. These contacts are broken when the complex transforms to the open complex. In the open complex the DNA is bent; this bending appears to be a key feature in the initiation of transcription. Additionally, the open complex is very stable, with a half-life measured in hours, compared to the closed complex, which has a half-life measured in minutes.

After the formation of the open complex, transcription can start. The first nucleotide of the transcript is nearly always a purine and binds within the β subunit. There appear to be two nucleotide binding sites within the β subunit, the first being the initiation site, and the second the polymerisation site.

3.4 Elongation

The transition from initiation to elongation is accompanied by major changes in the structure of the transcription complex. The complex becomes very stable and is resistant to dissociation. The σ subunit is released from the advancing polymerase after the addition of nine or ten nucleotides, leaving the core enzyme to complete the transcript.

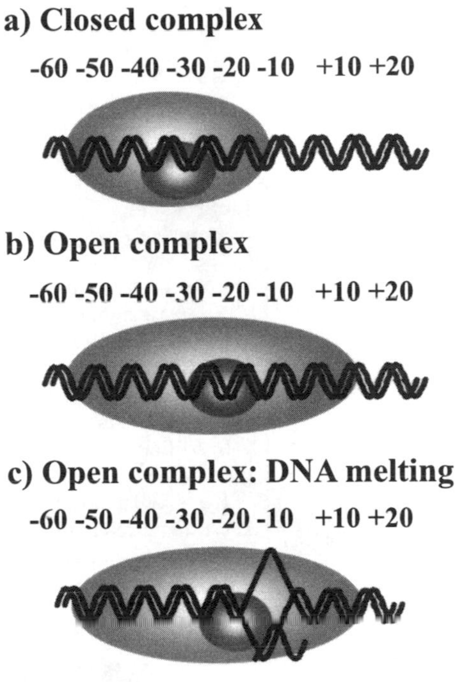

Figure 5.1 Open and closed complexes

DNA footprinting – not to be confused with DNA fingerprinting – is a powerful technique for studying the binding of proteins to DNA.

Box 5 DNA Footprinting

DNA footprinting is a powerful technique used to assess ligand binding to specific regions of DNA. It is separate and distinct and not to be confused with the much discussed technique of DNA *finger*printing, which involves analysis of variable length repeats and is used in forensic applications, especially in murder, rape, and paternity investigations.

In footprinting, the DNA fragment is radiolabelled at one end of one strand, by a kinase reaction using γ-^{32}P-dATP or by filling in a sticky end using the Klenow fragment of DNA polymerase and α-^{32}P-dATP. After labelling, the DNA is then partially digested using a specific nuclease (usually DNase I) in the presence and absence of the ligand of interest. Typically this ligand is a DNA-binding drug, other oligonucleotide, or in the study of transcription, more usually a DNA-binding protein. The products of digestion are then separated on a high-resolution, denaturing, polyacrylamide gel, which is then dried onto filter paper and exposed to x-ray film.

A ladder of bands is observed in the control lane, corresponding to partial digestion of the DNA by the nuclease in the absence of the ligand. In the presence of the ligand, the target site is protected from digestion, revealing a gap or "footprint" amongst the ladder of bands. The position of this recognition site within the DNA fragment can be determined by running marker lanes of DNA fragments of known sizes alongside the digestion products, and can be confirmed by comparison of the sequence and the sizes.

DNase I as a Footprinting Agent

DNase I is the major enzyme used in footprinting; it is a nuclease isolated from bovine pancreas and has a molecular weight of 30.4 kDa. The enzyme cuts double-stranded DNA, introducing single-strand breaks into the

Box 5 *continued*
DNA
Footprinting

phosphodiester backbone by hydrolysis of the O-3′-P bond. The enzyme requires divalent cations exhibiting optimal activity in the presence of Ca^{2+}, Mg^{2+}, or Mn^{2+}.

Although DNase I cleavage is not sequence-specific, the enzyme does have a slight preference for the sequences 5′-TN, 5′-AN, 5′-NT, and 5′-NA, where "N" can be any base. Additionally, DNase I cleavage can be correlated with local and global variations in DNA structure. These results have been confirmed by the determination of the structure of several complexes between oligonucleotides and DNase I.

The enzyme acts by inserting an exposed loop into the minor groove of the DNA helix. This mechanism of action explains why DNA regions with unusually narrow minor grooves, such as polydA.polydT, are poor substrates.

In the crystal structures reported, the DNA is bent towards the major groove, away from the enzyme. If this bending is a necessary part of the enzyme's catalytic mechanism then it may explain why GC-rich regions, which are expected to be more rigid, are also typically resistant to DNase I cleavage.

In addition it has been suggested that DNase I cleavage may be sterically hindered by the presence of a guanine two to three bases from the cutting site of the enzyme. Local variations in helix twist angle, base pair propeller twist, and tilt and roll angles also affect the DNase I-DNA interaction.

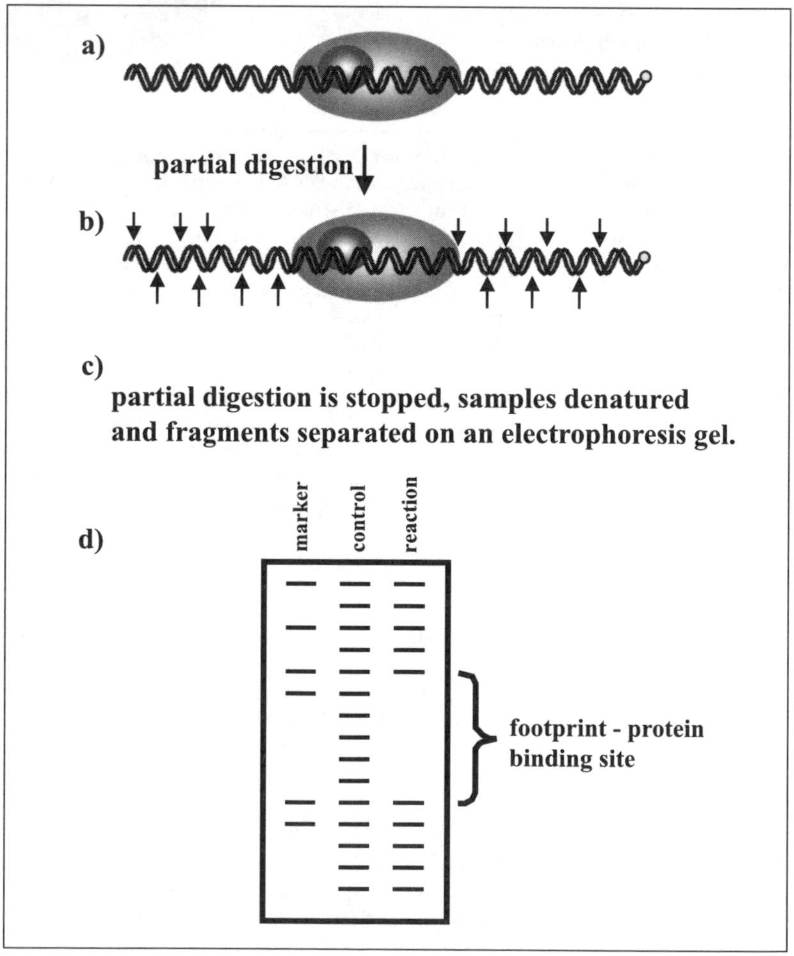

DNA footprinting. The DNA binding site of a specific protein can be mapped using the footprinting technique. The DNA is artificially labelled on one strand.

a) The protein of interest is mixed with the DNA and allowed to bind.

b) A DNA-digesting enzyme (typically DNase I) is then added and allowed to partially digest the DNA. The section of the DNA bound by the protein will be protected from digestion.

c) The digestion is stopped. The DNA is then denatured and the fragments are separated on a denaturing electrophoresis gel.

d) By analysing the banding pattern relative to the control lane and the marker lane (containing DNA with no protein bond), it is possible to identify the binding site of the protein by the missing bands, the "footprint" of the protein.

Major and minor grooves of DNA. The DNA double helix contains two grooves, referred to as the major and minor grooves.

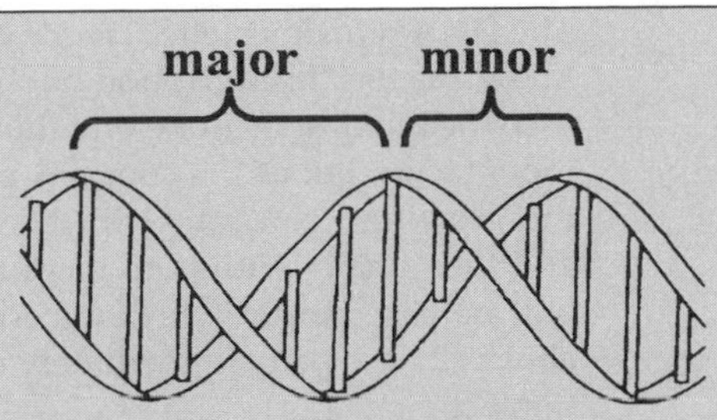

Box 5 *continued*

Hydroxyl Radicals as a Footprinting Agent

Hydroxyl radical DNA cleavage is far less sequence selective than DNase I. Hydroxyl radicals are generated by the reduction of hydrogen peroxide by ferrous ions. The hydroxyl radicals cut the DNA duplex randomly, along its length. Hydroxyl radicals have been used in connection with the study of nucleosomal DNA (see chapter 7).

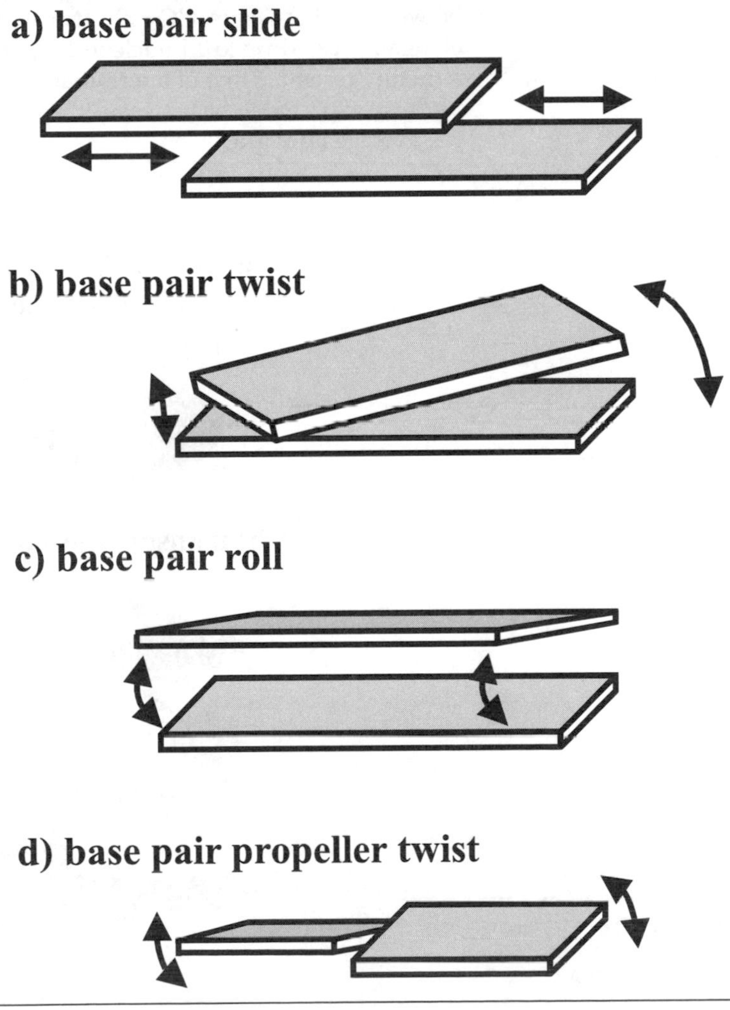

DNA structure. Movement is possible between bases and between strands in DNA, allowing for local variations in structure.

The DNA remains melted over a region of about 17 bases in a structure known as the "transcription bubble" (Figure 5.2).

In the initial stages of elongation, the RNA is in close proximity with the σ factor, but as the complex proceeds and the σ factor dissociates, the growing RNA transcript becomes associated with the β subunit and then finally with both the β and β' subunits. The RNA produced will be base paired with the coding DNA strand within a region of about 12 base pairs immediately upstream of the point of polymerisation activity.

The prokaryotic RNA polymerase has proofreading capacity, which resides in the β subunit.

The prokaryotic RNA polymerase appears to have some proofreading ability and this seems to reside within the β subunit. In the wild-type *E. coli* the error rate is approximately one base per million transcribed. Mutant bacteria are known which lack this proofreading ability; the error rate can rise to one in one hundred bases and the speed of elongation is increased.

The rate of elongation will vary depending on the gene itself, especially the presence of GC-rich sequences. More energy is needed to melt the three hydrogen bonds between GC base pairs than the two between AT base pairs. GC-rich sites are referred to as "pause sites" – elongation can drop to 0.1 nucleotides per second. These sites and their structure resemble that of a termination region. Maximum elongation rates are expected to be between 30 and 60 bases per second, although the average rates are lower.

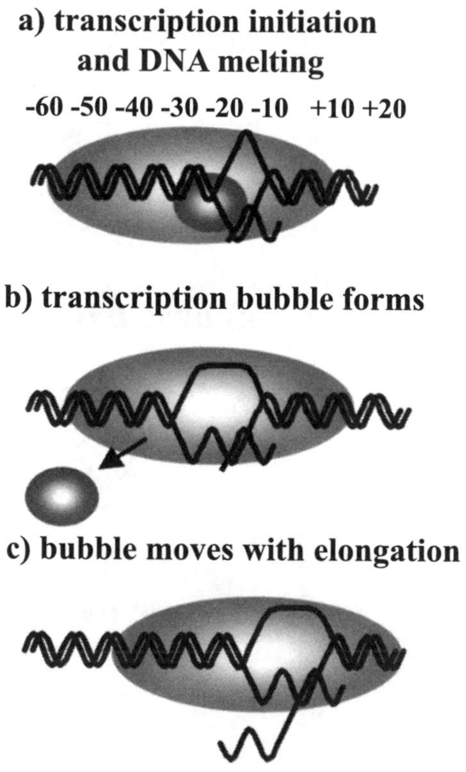

Figure 5.2 The transcription "bubble", a region of nuclease-sensitive, single-stranded DNA.

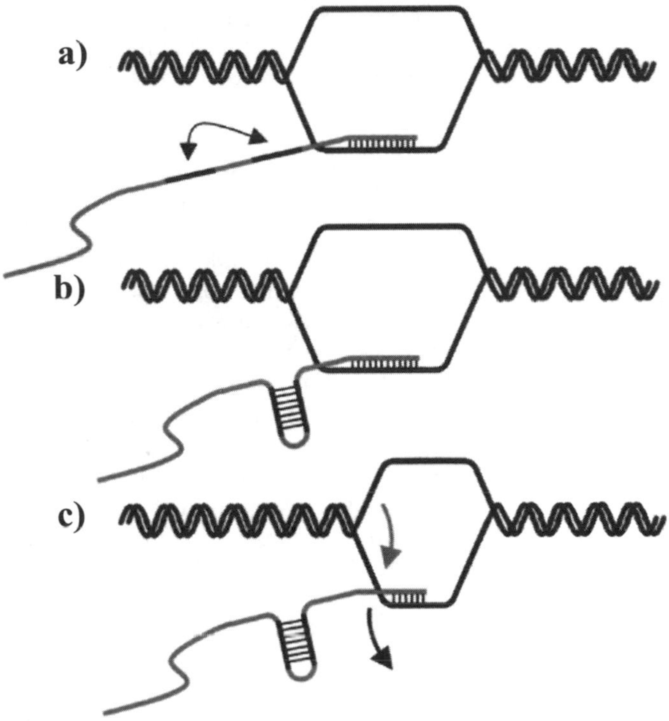

Figure 5.3 Rho (ρ)-independent termination.

3.5 Termination

In *E. coli* two types of termination mechanisms have been identified. The first uses a signal sequence and requires no additional protein factor. This mechanism operates in the ribosomal RNA genes. The second mechanism requires an additional protein factor known as rho (ρ).

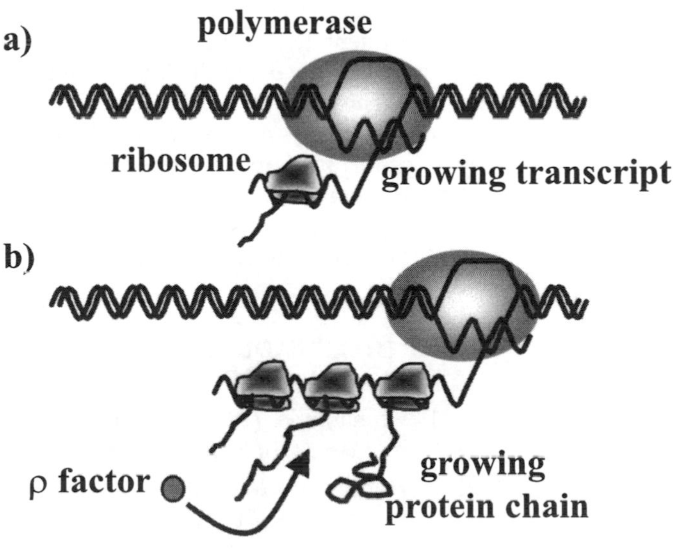

Figure 5.4 Blocking of rho (ρ) factor binding.

In the case of signal sequence recognition, the transcribed mRNA forms a stem loop structure. This structure will cause the RNA polymerase to slow down and pause and the subsequent incorporation of a run of uridine (U) residues forms base pairs with a run of A residues contained in the DNA template. The A–U base pair is weak and is not strong enough to hold the RNA product, allowing the RNA to be released. The RNA polymerase is then released from the DNA template (Figure 5.3).

The second termination mechanism requires the rho (ρ) protein. The ρ factor is an essential protein for the normal growth of *E. coli*. It exists as a hexamer *in vivo* and is an enzyme with ATPase activity. This ATPase activity is only functional in the presence of single-stranded RNA more than 80 bases in length. The ρ factor binds and causes transcription termination and release of the RNA. In genes dependent on the ρ factor, there is no transcription stop signal; the polymerase simply continues until the ρ factor binds the transcript and causes termination.

The obvious question then is why does the ρ factor not cause premature termination of all transcripts of 80 base pairs? The answer seems to be that in prokaryotes, ribosomes bind to the nascent RNA and begin to translate the RNA into proteins.

These ribosomes working along the mRNA block the binding of the ρ factor until the RNA transcript has passed the ribosome stop codon, contained within the RNA sequence (Figure 5.4).

Summary

The bacterial RNA polymerase has been extensively studied both for its own sake and as a model for the more complex eukaryotic enzyme. The α, β, and β' subunits assemble and this complex is transcriptionally active, or rather it can be when it locates a promoter, with the assistance of the σ factor. When the polymerase binds to a promoter, significant changes occur in it and the DNA, giving a transition from the closed to the open complex. This structural change can be detected experimentally, using, for example, DNA footprinting. Transcription then begins and the RNA polymerase complex moves off down the gene. The σ factor then dissociates. There are two mechanisms of termination, one involving a signal sequence that leads to the formation of a hairpin loop, the other involving another protein, ρ (rho). In prokaryotes, there is virtually no RNA processing and ribosomes bind to the mRNA almost as soon as it emerges from the RNA polymerase and immediately begin to translate it.

Further reading

1. Ishihama, A. (1988) Promoter selectivity of prokaryotic RNA polymerases. *Trends Genet.* **4**, 282–286.
2. Henkin, T. M. (1996) Control of transcription termination in prokaryotes. *Annu. Rev. Genet.* **30**, 35–57.

3. Haldenwang, W. G. (1995) The sigma factors of *Bacillus subtilis*. *Microbiol. Rev.* **59**, 1–30.
4. Ishihama, A. (1990) Molecular assembly and functional modulation of *Escherichia coli* RNA polymerase. *Adv. Biophys.* **26**, 19–31.
5. Pugh, B. F. (1996) Mechanisms of transcription complex assembly. *Curr. Opin. Cell Biol.* **8**, 303–311.
6. Narberhaus, F. (1999) Negative regulation of bacterial heat shock genes. *Mol. Microbiol.* **31**, 1–8.
7. Yanofsky, C., Konan, K. V. and Sarsero, J. P. (1996) Some novel transcription attenuation mechanisms used by bacteria. *Biochimie.* **78**, 1017–1024.

END OF UNIT QUESTIONS

1. Which subunits are necessary for a transcriptionally active bacterial RNA polymerase and in what order do they assemble?
2. Why does the ρ protein not cause premature termination of all transcripts?
3. What is a transcription bubble and how might it be studied experimentally?

PROBLEMS

1. Discuss the evidence for the ordered assembly of RNA polymerase components (see references 4 and 5).
2. Discuss the role of the sigma (σ) factor in bacterial transcription (see reference 3).

6 Eukaryotic Transcription

Objectives

a) introduce the three eukaryotic RNA polymerases, with an emphasis on RNA polymerase II

1. Overview

Having seen the prokaryotic RNA polymerase, we now move to eukaryotes. Eukaryotes have three different, though related, RNA polymerases, which transcribe RNA from different classes of genes. Additionally, the eukaryotic genome is much larger and as we have already seen (chapter 4), promoters are larger and more complex. Also, in eukaryotes, there is considerable post-transcriptional RNA processing (chapter 2).

2. Transcription in the eukaryotic nucleus

Prokaryotes such as *E. coli* contain a single type of RNA polymerase that produces all of the cell's RNA; in eukaryotes, however, there are three. The three types of RNA polymerase differ in template specificity, localisation, and susceptibility to inhibitors.

Eukaryotes have three different, though related, RNA polymerases, which transcribe different classes of genes.

The three RNA polymerases (I, II, and III) in eukaryotic cells each comprise many subunits. The so-called "core" RNA polymerase subunits have no intrinsic capability to transcribe the DNA template accurately to produce RNA, even at a basal level. Each RNA polymerase requires a distinct group of general accessory factors to provide selectivity in their actions and to allow accurate transcription to occur.

Genes transcribed by RNA polymerases I and III are believed to have DNA-bound gene-specific activators "resident" at (*i.e.*, permanently bound to) their promoters. This may not be the case for RNA polymerase II.

2.1 RNA polymerase I

RNA polymerase I is located in the nucleolus (a distinct structure within the nucleus) where it transcribes only the tandem array of genes for the ribosomal RNAs (rRNAs). It is insensitive to the drug α-amanitin. The genes for these rRNAs are clustered together and initially are transcribed as a single 45S RNA which is then enzymatically modified and cleaved to yield the mature 18S, 5.8S, and 28S rRNAs (RNA polymerase III transcribes the fourth ribosomal RNA molecule).

RNA polymerase I transcribes ribosomal RNA genes.

RNA polymerase I promoters are characterised by the absence of a TATA box and the presence of two key sites, a GC-rich region, located at approximately –45 to +20, and a second GC-rich region, referred to as the upstream control element (UCE), typically at approximately –180 to –107.

2.2 RNA polymerase III

RNA polymerase III transcribes the 5S rRNA, tRNA, and snRNA (small nuclear RNA) genes. It is located in the nucleoplasm and is α-amanitin-sensitive in some species.

RNA polymerase III transcribes the 5S rRNA, tRNA, and snRNA genes.

Two types of promoter are known in RNA polymerase III genes; the promoter is downstream of the transcriptional start site in the 5S rRNA and tRNA genes and upstream of it in the snRNA genes. The 5S RNA gene promoter contains no TATA box and is located at +55 from (*i.e., downstream* of) the transcriptional start site (referred to as Box A). A second key sequence is located at +80 to +90 (referred to as Box C). TFIIIA binds the BoxA–BoxC region and appears to promote the binding of TFIIIC to the same area. Similarly, the tRNA genes contain two key sequences, both *downstream* of the transcription initiation site. Again, there is no TATA box.

A second type of RNA polymerase III promoter is seen in the snRNA genes. This is found *upstream* of the gene and is fairly variable among genes and different species. Such promoters contain binding sites for the OCT transcription factor, a proximal sequence element (PSE), and a TATA box located close to the PSE.

2.3 RNA polymerase II

Messenger RNA (mRNA) precursors are synthesised by RNA polymerase II.

Messenger RNA (mRNA) precursors are synthesised by RNA polymerase II, which is also located in the nucleoplasm. This enzyme transcribes all of the cell's complement of mRNA molecules and several small RNA molecules, such as the U1 snRNA used in the splicing apparatus. RNA polymerase II is α-amanitin sensitive. The full enzyme comprises 12 subunits.

RNA polymerase II promoters typically contain a TATA box, 15–25 nucleotides upstream of the transcription start site, although TATA-less promoters are known. Additionally, the sequence Py_2CAPy_5 is usually seen at the transcriptional initiation site, the A being position +1 or the first base of the transcript. Often, RNA polymerase II promoters also contain a GC-box centred at around –90. The consensus sequence is 5′-GGGCGG and often promoters contain more than one copy. This sequence is a binding site for the transcription factor Sp1 (stimulatory protein 1). The so-called "CCAAT box" is also a common element in RNA polymerase II promoters.

RNA polymerase II transcribes messenger RNA, in conjunction with a multitude of other proteins. Some of these are essential to ensure the mechanics of the transcription process itself (so-called *general* or *basal* transcription factors) and others are required to select the genes to be transcribed at any given moment or in a given cell type and regulate the rate of their transcription (the *activator* and *repressor* factors). This selection of genes to be transcribed is based largely on

Figure 6.1 Common elements in promoters of genes transcribed by RNA polymerase II

Element	Consensus Sequence	Transcription Factor that binds
TATA box	TATAAAA	TBP (TATA binding protein)
CAAT box	GGCCAATCT	CTF/NFI binding factor/ necrosis Factor 1)
GC box	GGGCGG	Sp1 (stimulatory protein 1)

specific recognition by these factors of certain sequences in the DNA surrounding the gene, especially the DNA upstream (5′ of) the transcriptional start site, that is, the promoter. A third class of factors contains proteins that act as links between activator/repressor proteins and basal transcription factors, transmitting gene-specific messages to the general transcription machinery. Such proteins are referred to as *co-factors*. In some cases co-factor proteins may be built-in components of the general transcription machinery. An example is TFIID, which comprises the TATA-binding protein (TBP) and many TBP-associated factors (TAFs). While TBP binds promoters at a specific site (the TATA box) and allows basal transcription at a low level, the modulation of transcription by activator proteins depends on the availability of TAFs.

Genes transcribed by RNA polymerase II typically contain common core-promoter sequences that are recognised by general transcription initiation factors and gene-specific DNA sequences that are recognised by regulatory factors. These, in turn, will modulate the function of the general initiation factors.

2.4 Transcription factories

Transcription occurs in discrete locations within the nucleus, not diffusely throughout it.

It was believed that RNA synthesis involved an RNA polymerase tracking along a static template searching out promoter sequences. However, recent research on chromatin structure suggests that the template DNA slides past a polymerase immobilised in a large transcription "factory" comprising many proteins. Consistent with this idea, immunocytochemical data suggests that sites of transcription are not randomly spread throughout the nucleus but are concentrated in discrete sites, which have been termed "transcription factories". Similarly, splicing factors are also located in discrete sites. What is still not clear is whether these sites, for transcription and splicing, are coincident, that is, whether transcription and splicing occur in the same place. In such a model, gene activation would involve reducing the gene-to-factory distance and increasing the affinity of a promoter for a factory.

Enhancers and similar control structures would attach to a factory and increase the chances that a promoter could bind to a polymerase; after transcriptional termination, the gene would detach from the factory. As some RNA processing (splicing, capping, tailing; chapter 2) occurs co-transcriptionally, processing sites are also likely to be associated with the factory.

Each chromosome in the haploid set has a unique array of transcription units strung along its length. Therefore, each chromatin fibre will be folded into a unique array of loops associated with clusters of polymerases and transcription factors; only homologs share similar arrays. As these loops and clusters, or transcription factories, move continually, they make and break contact with others. Correct pairing would be nucleated when a promoter in a loop tethered to one factory binds to a homologous polymerising site in another factory, before transcription stabilises the association.

2.5 General transcription factors

As in prokaryotes, eukaryotic transcription occurs in three phases: initiation, elongation, and termination.

Transcription in eukaryotic cells shares many features with the process in prokaryotes. As mentioned before, eukaryotic cells have three distinct RNA polymerases.

Similar to the scenario observed in prokaryotes, eukaryotic transcription can be separated into several discrete processes. First, the so-called pre-initiation complex (PIC) forms, and then becomes "activated", a process involving DNA melting. When this has occurred, transcription can be initiated. Next, promoter clearance occurs, that is, the polymerase moves away from the promoter and continues to extend the mRNA. Finally, there is strand termination. The major difference in these processes between eukaryotes and prokaryotes is the greater complexity in eukaryotic cells.

2.5.1 *The component parts of RNA polymerase II*

RNA polymerase II comprises 12 subunits.

RNA polymerase II comprises 12 subunits, but even this is insufficient to ensure accurate transcription; additional factors are needed even at the strongest core promoters. These transcription factors are often themselves complex, comprising several subunits. Together they are functionally equivalent to the single σ-factor in prokaryotes. In prokaryotes the σ-factor is directly responsible for *specific* gene expression.

However, in eukaryotes the only transcription complex that can direct sequence-specific transcription initiation is TFIID. The components of TFIID include the TATA-binding protein (TBP) that binds to the TATA box contained within the promoters of the gene and associates with other TBP-associated factors (TAFs). The TAFs have been directly implicated in sequence-specific DNA binding.

Although the concept of pre-formed holoenzyme complex being able to perform efficient transcription is gaining acceptance, the prevailing idea is that the pre-initiation complex is assembled in an ordered, stepwise fashion. This will now be described.

2.5.2 *The pre-initiation complex*

The fundamental complexes of the pre-initiation complex (PIC) assembly, the so-called "general" factors, include TFIID, TFIIB, TFIIE, TFIIF, TFIIH, and RNA polymerase II (Figure 6.2).

Initially, TFIID binds to the TATA sequence in the promoter, an interaction mediated via the DNA minor groove. This complex is stable and distorts the DNA helix somewhat, bringing sequences upstream and downstream of the TATA box into closer proximity.

TFIID itself comprises the TATA box binding protein (TBP) and TBP-associated factors (TAF_{II}s). The name of the major component of TFIID, "TATA box binding protein" (TBP), is somewhat unfortunate and misleading, because it appears not only to bind to the TATA box, but to other sequences, and is required by all three eukaryotic RNA polymerases, not just RNA polymerase II. It is an essential component of both the initiation complex "SL1", formed on the TATA-less ribosomal RNA promoters that are transcribed by RNA polymerase I,

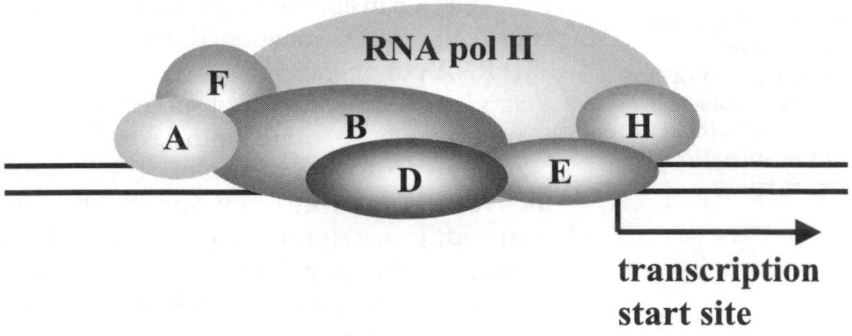

Figure 6.2
Pre-initiation complex.

and the TFIIIB complex formed on some RNA polymerase III promoters. Despite this, the name has stuck.

TBP is a highly conserved protein and contains two DNA-binding direct repeats each of 66 amino acids, each flanked by a basic region. Its shape is believed to be saddle-like, and the protein sits over the TATA box, interacting with the DNA minor groove. This binding induces a considerable bend into the DNA, which seems to be an important component of the site-specific recognition by TBP. TBP contacts RNA polymerase II via the carboxyl-terminal domain of its largest subunit.

The binding of TFIID to DNA allows TFIIA to bind to the complex through direct contacts with TBP and upstream DNA sequences. Such binding of TFIIA gives the complex extra stability.

The next component to bind is TFIIB, again further stabilising the growing complex. The binding of TFIIB is mediated via direct interactions with TBP and with DNA sequences upstream and downstream of the TATA box. However, TFIIB's primary role is in transcriptional start site selection.

From x-ray crystallographic studies, it has been shown that the binding sites for TFIIA and TFIIB do not overlap and that therefore the two proteins can bind simultaneously. Additionally, TFIIA and TFIIB have no direct contacts with each other. TBIIB is involved directly in RNA polymerase II-TFIIF recruitment, whereas TFIIA has an anti-repression function where it is involved in the dissociation of negative co-factors from the TBP complex, which otherwise would prevent TFIIB binding to the PIC.

TFIIF and RNA polymerase II then interact with each other, forming a stable complex. This complex will bind to pre-formed PIC through direct interactions with TFIIB. Although TFIIF has a direct role in promoter targeting by RNA polymerase II through these interactions, it also plays an indirect role by reducing RNA polymerase II binding to non-specific sites in DNA. TFIIF also plays a dual role in that it serves as a transcription elongation factor, though the domains implicated in this function are different from those involved in initiation.

The next factor, TFIIE, binds through direct interactions with RNA polymerase II and may bind to TFIIF and TBP. One of the subunits of

X-ray crystallography is a major technique for determining the three-dimensional structure of DNA, proteins, and DNA-protein complexes.

Box 6 Determination of the Three-Dimensional Structures of DNA, Proteins, and DNA-Protein Complexes

Both x-ray crystallography and nuclear magnetic resonance (NMR) spectroscopy have been used to study complexes of general transcription factors and transcriptional activators with their DNA targets. Such studies have provided important structural insights into transcription initiation by RNA polymerase and DNA sequence recognition.

X-ray Crystallography

X-ray crystallography is the most widely used experimental technique for the precise determination of the three-dimensional structures of large and small molecular weight compounds. Advances in instrumentation and computational methods have greatly facilitated the procedure and the usual bottleneck now is obtaining crystals of suitable size and quality.

X-rays having an appropriate wavelength (~10^{-10} m) are scattered by the electron cloud of an atom. Based on the pattern of diffraction caused by the regular, repeating assembly of molecules or atoms in the crystal, a map of the electron density - and therefore atomic positions - can be constructed. When combined with sequence and other biochemical and biophysical information, an accurate molecular model can be developed. Such models have been successfully used in establishing structure–function relationships and provide a basis for application-oriented research in protein design and drug design.

Several three-dimensional crystal structures of huge importance to the study of transcription have been determined in recent years. These include TATA-binding protein (TBP), TBP bound to a TATA DNA sequence, and the ternary complex of TFIIB binding TBP bound to a TATA element.

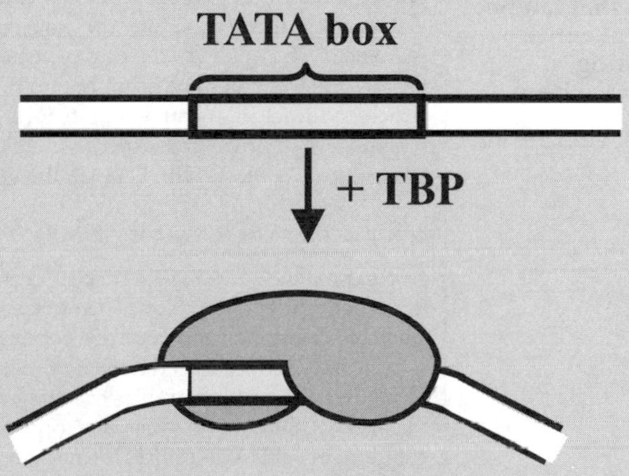

TBP is a highly symmetric α/β structure with a DNA-binding fold, which resembles a molecular "saddle", sitting astride the DNA. The DNA-binding surface seems to be a novel curved, anti-parallel β-sheet.

The structure of TBP complexed with a TATA element revealed a dramatic conformational change in the DNA, causing a sharp kink at both ends of the 5′-TATAAAAG sequence. Between the kinks, the right handed DNA double helix is curved and slightly unwound, producing a widened minor groove that interacts with TBP's concave, anti-parallel β-sheet.

The TFIIB/TBP/TATA ternary complex structure reveals that core TFIIB recognises the pre-formed TBP-DNA complex via protein–protein and protein–DNA interactions.

As is discussed in chapter 4, NF-κB is the archetypal member of the Rel family of transcription factors. NF-κB and other members of the family play diverse roles in regulating the immune response, cell growth, and development. NF-κB and the other Rel family members share an amino-terminal Rel-homology domain (RHD). This region of the proteins is responsible for dimerisation, DNA binding, and, when complexed with IκB, the cytosolic localisation of unactivated protein. Signal-induced activation causes the dissociation of the NF-κB–IκB complex and the translocation of the active transcription factor into the nucleus.

NF-κB binds to κB sequence elements, the consensus sequence of which is 5′–GGGPuNPyPyPyCC–3′, where "Pu" is a purine, "Py" is a pyrimidine, and "N" is any nucleotide. The crystal structure NF-κB (the p50/p65 heterodimer) bound to such a κB DNA element has been determined. This structure shows that the p50 and p65 subunits of the heterodimer each bind DNA, there being a 5-base 5′ subsite for p50 and a 4-base 3′ subsite for p65.

Another DNA-binding structure is that of the "T-box". The mouse Brachyury (T) gene is the prototypical member of the T-box gene family. Mutations in T-box genes often lead to drastic embryonic phenotypes, pointing to the essential role of the gene products in tissue specification, morphogenesis and organogenesis.

NMR is a technique for determining solution structures of biomolecules. It is currently somewhat limited in the size of molecules that can be studied, but entire DNA-binding domains have been examined.

The T-box encodes the DNA-binding T domain, which is approximately 180 amino acids in length. The crystal structure of a T-domain protein complexed with a 24-base palindromic DNA sequence has been reported. The proteins bind DNA as a dimer and interact with both the major and the minor grooves. The structure also revealed a novel type of DNA–protein contact, in which an α helix and the carboxyl terminus of the protein is embedded within the minor groove without causing the DNA to bend.

Proteins in the fos and jun families form heterodimers and bind to the DNA sequence 5′-TGAGTCA. The crystal structure of a heterodimer comprising the bZIP regions of c-Fos and c-Jun bound to DNA has been determined. The structure reveals that both subunits form α-helices. The carboxyl-terminal regions form a coiled-coil structure, while the amino-terminal regions make base-specific contacts with DNA in the major groove.

Nuclear Magnetic Resonance (NMR)

Nuclear magnetic resonance spectroscopy (NMR) is another major technique for structural determination. The nuclei of all atoms carry a charge; when the spins of the protons and neutrons comprising these nuclei are not equal, the overall spin of the charged nucleus generates a magnetic dipole, which can be measured. Common examples of atoms with such a dipole include ^{1}H, ^{13}C, ^{15}N, ^{19}F, and ^{31}P. Obviously, with the exception of ^{19}F all these atoms are common components of biomolecules, including DNA and proteins.

The yeast transcription factor ADR1 binds DNA via two cys_2-his_2 (C_2H_2) zinc fingers. The minimal DNA-binding domain comprises these zinc fingers and 20 amino acids proximal and amino-terminal to them.

The structure of this DNA-binding domain has been studied using proton, ^{15}N, and ^{13}C NMR. The analysis suggests that the two fingers bind DNA with different orientations, as the entire helix of finger 1 is perturbed, while only the extreme amino-terminal region of the finger 2 helix is affected. The residues identified as being key in making contact with DNA were previously shown by mutagenesis experiments to be important.

Box 6 *continued*

the TFIIE complex has been shown to interact with the DNA region just upstream of the start site. This is consistent with the expected function of TFIIE, that of assisting in bringing about DNA melting.

The last factor to bind, TFIIH, completes the assembly of the PIC. TFIIH is believed to interact directly with TFIIE; consistent with this, TFIIH apparently stabilises TFIIE binding (Figure 6.3).

2.5.3 *Co-factors*

During early studies of transcription it was thought that positive regulatory proteins (or activators) binding to the enhancer element exerted their effects directly, through physical contact with components of the transcription complex. This may be the case in some instances, but it is now believed that the effects of some activators are indirect, via intermediary factors. The failure of the systems reconstituted with purified transcription subunits to transcribe mRNA

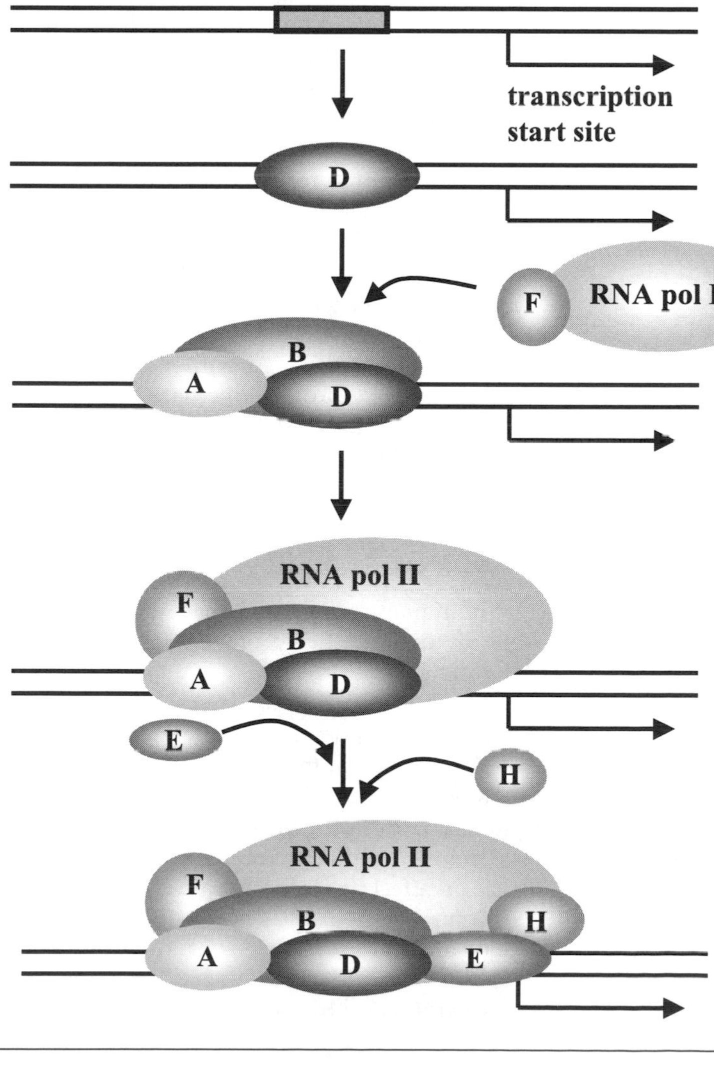

Figure 6.3 Assembly of the pre-initiation complex.

indicated that specific factors were missing. Through cell fractionation some of these factors have since been identified. To date, three types of intermediates have been described: (1) TATA-binding protein associated factors (TAFs) which associate with TATA-binding protein (TBP) to form TFIID, (2) mediators, which associate with RNA polymerase II to form a holoenzyme, and (3) general co-factors.

2.5.4 *TATA-binding protein (TBP)-associated factors (TAFs)*

The TATA-binding protein (TBP)-associated factors (TAFs) are subunits of TFIID and are thought to perform critical functions in both transcription activation and core-promoter recognition.

TFIID appears to be the only general transcription factor to have the ability to bind to DNA in a sequence-specific manner. It is also the first subunit of the pre-initiation complex to associate with the DNA, where it initiates the recruitment of RNA polymerase II and other general transcription factors.

After the initial cloning of the TATA-binding factor (part of TFIID) it was noted that this could replace the TFIID complex in basal transcription. However, this setup would not support activated transcription. This result led to the conclusion that there was something missing and it was hypothesised that the TFIID fraction must contain additional subunits that are *dispensable* for basal transcription, but are *essential* for regulated transcription. The missing pieces of TFIID were discovered in *Drosophila* and have since been identified in human cells. TFIID from human and *Drosophila* cells is a multi-protein complex comprising TBP and at least eight associated subunits, referred to as the "TATA-binding protein (TBP)-associated factors (TAFs)".

The TAFs associated with TBP act to help activate transcription. How do these proteins mediate their effect(s)? Perhaps the simplest interpretation would be that interactions between the activator protein TAFs lead to increased recruitment of TFIID to the core promoter.

2.5.5 *General co-factors*

Within mammalian cells, there are many co-factors. These can broadly be divided into two groups, positive co-factors (PCs) and negative co-factors (NCs).

An example of a positive co-factor is PC4. The gene for this protein has been cloned and the structure of the protein studied. PC4 interacts with the general transcription factor TFIIA. TFIIA accelerates and stabilises binding of TFIID to the TATA box. This can be shown by "functional order-of-addition" experiments in purified transcription systems. If TFIID is already bound to the TATA box, addition of PC4 has no effect on transcription.

Interaction of PC4 with TFIIA and transcriptional activators indicates that it functions as a bridge between the activator and the basal transcription machinery. It is possible that PCs strengthen direct activator–basal transcription machinery interactions.

An example of a negative co-factor is NC2. This protein acts by forming a stable complex with TBP and can then repress basal transcription. NC2's mechanism of action is to compete with the binding of TFIIB to TBP, thus inhibiting the assembly of the initiation complex. The NC2 consists of two subunits (NC2α and NC2β). The sequence of the NC2 complex displays strong homology to the histones H2A and H2B within a region that is conserved in all four histones. This region is known as the histone fold motif (discussed below). Transcription inhibition by NC2 is dependent on this motif, which is responsible for the dimerisation of the two subunits of the NC2 complex. NC2's role in the cell is to control overall basal transcription activity. This occurs through inhibiting the actions of TFIIB. NC2 is strongly conserved from yeast to humans, pointing to a key role in transcription.

2.6 Transcription elongation

For transcription to continue after the complex clears the promoter, the proteins surrounding RNA polymerase rearrange. The factors needed for initiation and promoter clearance leave the complex and are replaced by elongation factors. Eukaryotes appear to have two classes of elongation factors: *general* elongation factors, the role of which is to promote efficient transcription for all protein-encoding genes, and *regulatory* elongation factors, which only participate in the control of transcription of specific genes or gene families. However, to date, no cellular proteins have been characterised the function of which clearly fits the role of gene-specific elongation factors.

To date, five general elongation factors have been described. These have been termed P-TEFb, SII, TFIIF, Elongin (SIII), and ELL.

2.7 Preventing arrest

P-TEFb and SII are structurally unrelated but have both been shown to protect RNA polymerase II from transcriptional arrest. How P-TEFb stops transcriptional arrest remains unclear. It is a heterodimer, comprising two proteins with molecular weights of approximately 124 kDa and 43 kDa. P-TEFb possesses protein kinase activity that can phosphorylate RNA polymerase II's carboxyl-terminal domain. Thus, it is possible that P-TEFb promotes elongation by a mechanism involving phosphorylation of RNA polymerase II and/or other proteins involved in transcription.

SII, a protein of approximately 38 kDa, seems to assist passage of RNA polymerase II through transcriptional impediments, including some nucleoprotein complexes and DNA sequences referred to as "intrinsic arrest sites". Such sites are located throughout transcribed regions of various eukaryotic genes.

2.8 SII and nascent transcript cleavage

Evidence suggests that elongation by RNA polymerase II proceeds by one of two mechanisms, referred to as "monotonic" and "discontinuous" elongation. During monotonic elongation, RNA polymerase II moves along the DNA each time a ribonucleotide is added to the growing transcript. When RNA polymerase II approaches an intrinsic arrest site or other potential blockage the enzyme will switch to discontinuous elongation and ceases to translocate even though the transcript continues to grow. The conformation of the polymerase will become strained. This can be resolved in one of two ways. Either the leading edge of the enzyme can move forward to escape arrest and resume monotonic elongation, or the catalytic site of the enzyme can slip backwards, out of contact with the 3′-hydroxy terminus of the nascent transcript, to enter an arrested state.

It is believed that SII can release RNA polymerase II from arrest by activating a polymerase-associated endoribonuclease, which cleaves the RNA transcript upstream of its 3′-hydroxy terminus, resulting in the creation of a new 3′-hydroxy terminus. This will be correctly positioned with respect to the polymerase catalytic site, which can now be extended.

2.9 Elongation factors that suppress pausing

The elongation factors TFIIF, Elongin (SIII), and ELL all increase the rate of elongation by suppressing pausing of RNA polymerase II. Like SII, they bind to RNA polymerase. Their mechanism of action remains unclear.

2.9.1 TFIIF

TFIIF is unique in eukaryotic transcription because it has a dual function, playing a role in both transcription initiation and elongation. Although TFIIF cannot release RNA polymerase II from an arrested state, it can decrease the probability of the arrest if it binds to the complex before it arrives at an SII-sensitive arrest site. TFIIF and SII appear to work to complement each other. TFIIF decreases the duration of pausing by RNA polymerase II and protects the elongation complex from arrest while SII re-activates elongation complexes that have arrested. It is unclear how TFIIF promotes elongation, though structure-function studies suggest that the protein's roles in initiation and elongation are mediated by different functional domains.

2.9.2 The elongin (SIII) complex

The elongin (SIII) multi-subunit complex comprises three subunits: A, B, and C. Subunit A is the largest (110 kDa) and has been shown to be the transcriptionally active subunit of the complex. Subunits B and C are much smaller, 18 and 15 kDa respectively. Subunits B and C can form a stable dimer, which can strongly induce elongin A transcriptional activity.

Summary

Eukaryotic transcription has much in common with that seen in prokaryotes. An obvious difference is the complexity; there are three related eukaryotic RNA polymerases, each comprising many subunits. The different RNA polymerases transcribe different genes; they recognise different promoter elements. RNA polymerase I is located in the nucleolus and transcribes only the tandem array of genes for the ribosomal RNAs (rRNAs). RNA polymerase III, which is found in the nucleoplasm rather than the nucleolus, transcribes the fourth ribosomal RNA molecule, the 5S RNA, and all the transfer RNA (tRNA) molecules. Messenger RNA (mRNA) precursors are synthesised by RNA polymerase II, as are several small RNA molecules, such as the U1 snRNA used in the splicing apparatus. Furthermore, many general transcription factors are also involved in forming a pre-initiation complex. Evidence suggests that transcription - and possibly splicing - occurs in "transcription factories", specific, discrete locations in the nucleus.

Further reading

1. Orphanides, G., Lagrange, T. and Reinberg, D. (1996) The general transcription factors of RNA polymerase II. *Genes Dev.* **10**, 2657–2683.
2. Svejstrup, J. Q., Vichi, P. and Egly, J. -M. (1996) The multiple roles of transcription/repair factor TFIIH. *Trends Biochem. Sci.* **21**, 346–350.
3. Roeder, R. G. (1996) The role of general initiation factors in transcription by RNA polymerase II. *Trends Biochem. Sci.* **21**, 327–335.
4. Zawel, L. and Reinberg, D. (1995) Common themes in assembly and function of eukaryotic transcription complexes. *Annu. Rev. Biochem.* **64**, 533–561.
5. Shastry, B. S. (1996) Transcription factor IIIA (TFIIIA) in the second decade. *J. Cell Sci.* **109**, 535–539.
6. Javier-Lopez, A. (1995) Developmental role of transcription factor isoforms generated by alternative splicing. *Dev. Biol.* **172**, 396–411.
7. Patikoglou, G. and Burley, S. K. (1997) Eukaryotic transcription factor-DNA complexes. *Annu. Rev. Biophys. Biomol. Struct.* **26**, 289–325.
8. Woychik, N. A. and Young, R. A. (1990) RNA polymerase II: subunit structure and function. *Trends Biochem. Sci.* **15**, 347–351.
9. Shilatifard, A. (1998) The RNA polymerase II general elongation complex. *Biol. Chem.* **379**, 27–31.
10. Geiduschek, E. P. and Tocchini-Valentini, G. P. (1988) Transcription by RNA polymerase III. *Annu. Rev. Biochem.* **57**, 873–914.

END OF UNIT QUESTIONS

1. List the types of genes transcribed by the three eukaryotic RNA polymerases and describe the common characteristics of the promoters in these types of genes.
2. Describe the nature and role of TFIID in eukaryotic transcription.

PROBLEMS

1. Discuss the evidence for the ordered assembly of the eukaryotic pre-initiation complex (see reference 4).
2. Discuss the subunit structure of eukaryotic RNA polymerase II. List the eukaryotic functional counterparts of the bacterial RNA polymerase subunits (see reference 8).

7 Chromatin and Eukaryotic DNA

Objectives

a) introduce the concept of chromatin and higher order DNA structure as an added complication in eukaryotic transcription

1. OVERVIEW

Eukaryotic transcription is not simply more complex because of the increased size of the genome and there being three different RNA polymerases. Eukaryotic DNA is tightly bound in higher order structures. Furthermore, many proteins are already bound to eukaryotic DNA before any of the RNA polymerases bind to a promoter. The nature of these proteins and how they influence transcription will be discussed in this chapter.

2. CHROMATIN AND TRANSCRIPTION

Eukaryotic DNA has many proteins already bound to it, which could interfere with the transcriptional process.

Eukaryotic cells contain approximately 2 metres of DNA, if stretched end to end. Obviously, given a cell the diameter of which is measured in microns, this DNA is very tightly packed. While this packing may be an extremely efficient way of squeezing the DNA molecules into a small cell, it is very problematic when the genome needs to be duplicated in mitosis or when a gene needs to be transcribed.

Much DNA is apparently simply inaccessible; a transcription initiation complex could not be formed. And yet it is. Electron micrographs demonstrate that transcriptional elongation does occur largely on DNA that is still packaged on nucleosomes. Additionally, *in vitro* at least, RNA polymerases can transcribe right through nucleosomal DNA.

2.1 The nucleosome

The basic unit of chromatin is the nucleosome.

A universal feature of chromatin packaging is the nucleosome, a structure in which 146 base pairs of DNA are wrapped approximately 1.7 times around an octamer of histone proteins, H2A, H2B, H3, and H4. This octamer is built from a $H3_2$-$H4_2$ heterotetramer, to which two heterodimers of H2A and H2B are joined (Figure 7.1). The DNA between nucleosomes is further constrained by linker histones (including H1 and H5) and by certain non-histone proteins, HGM 1 and HGM 2 being the most common. The importance of the histone proteins as regulators of gene expression has been known for some time. Formation of a DNA–nucleosome complex effectively represses the initiation of transcription; the compaction of DNA into chromatin presents many obstacles. Both nucleosomes themselves and the higher order structures into which they are arranged limit the access of *trans*-acting factors to their recognition sequences.

It was once thought that these histone and non-histone proteins simply packaged the cellular DNA into a highly condensed structure in order to physically fit all the DNA into the nucleus in an ordered and controlled way.

The most basic level of organisation of eukaryotic genomic DNA is the nucleosome, approximately 10 nm in diameter. When chromatin is isolated in a low salt buffer, it is possible to see a structure resembling beads on a string, there being approximately 150–200 base pairs of DNA wrapped around the histone octamer.

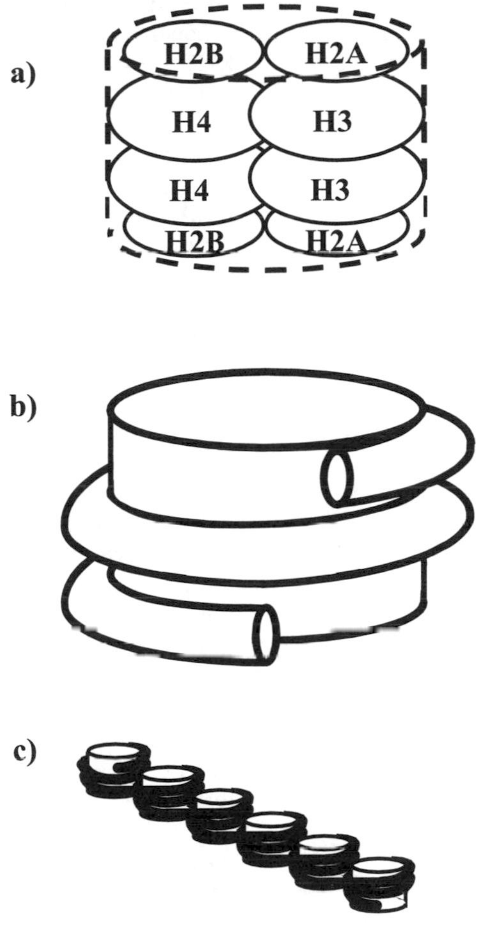

Figure 7.1 The elements of chromatin structure.

a) The basic structure of the nucleosome, indicating the relative positions of the individual proteins.
b) DNA wraps around the protein octamer (approximately 1.7 times) forming a left-handed superhelix.
c) The nucleosome cores form a repeating structure on the DNA strand, resembling beads on a string.

More recently, evidence demonstrates that nucleosomes play an important role in the regulation of eukaryotic gene expression. Specifically, using both *in vivo* and *in vitro* approaches it has been shown that nucleosomes can repress the transcriptional process, and that this inhibition can be overcome by positively acting transcriptional regulatory proteins. Studies using yeast mutants exhibiting a subnormal number of histones, because of low histone synthesis, show that genes in such cells can become constitutively active, even in the absence of their gene-specific activators.

2.2 Chromatin remodelling

Alterations in the chromatin structure occur at promoter and enhancer regions, either before or concurrent with the induction of transcription. These structural changes are often revealed by an increase in sensitivity at specific sites to digestion by the nuclease DNase I, and are as a result termed DNase I hypersensitive sites (DHSs). Most DHS sites are formed when protein activators interact with upstream promoter elements; presumably, these interactions cause remodelling of the chromatin structure. The formation of DHS sites at the regulatory regions appears to be a prerequisite for transcriptional initiation, and is therefore likely to represent an early step in the gene activation pathway.

Additionally, it has been shown that bacteriophage and eukaryotic RNA polymerases can transcribe through reconstituted nucleosomal DNA unless the promoter was "blocked" by a nucleosome actually bound to the promoter. However, it has also been shown that if the promoter DNA extended far enough beyond the nucleosome, RNA polymerases could start making a transcript and continue to extend it, regardless of the presence of nucleosomes bound further along in the sequence.

What interaction causes this remodelling of the chromatin and what does remodelling mean? How might the interactions of the transcriptional activators lead to the formation of DNase I hypersensitive sites?

Transcriptional activators can bind to their recognition sequence elements within nucleosomal DNA, leading to the formation of a ternary complex containing activators, histones, and DNA. As a consequence of transcription factor binding, histone-DNA contacts may be partially disrupted. This can lead to histone displacement from the transcription factor-bound sequences either in a *trans* (different DNA strand) or in a *cis* (same DNA strand) position. The resulting "nucleosome-free" gap in the DNA represents a DHS, because the loss of the histones would leave the DNA exposed and susceptible to DNase I digestion. Alternatively, the formation of the ternary complex may alter the nucleosome–DNA interaction but not lead to release of the nucleosome. The increased sensitivity to DNase I could be the result of an altered DNA structure. In principle then, a DHS could represent either a nucleosome-free region or an altered structure involving co-occupancy on the DNA of transcription factors and histones.

The nature of the nucleosome–DNA interaction may sterically hinder interactions of transcription factors with their target sites. However, *in vitro* studies have shown that many transcription factors can compete with histones for their binding sites. In such studies, nucleosomes inhibit the binding of activators, relative to naked DNA, from as little as two-fold to approximately one thousand-fold. The different DNA binding motifs contained within the trans-activator protein will also affect the efficiency with which the protein binds nucleosomal DNA relative to naked DNA.

A second determinant of transcription factor binding affinity is the location of the transcription factor binding site within the DNA on the nucleosome core. DNA does not bind to the histone proteins with the same affinity along the entire nucleosome core. Binding is known to be weaker where the DNA enters and leaves the nucleosome core, and so would be an "easier" target for transcription factor binding. Additionally, a model has been proposed whereby DNA sequences are transiently released from the surface of the histone octamer, thereby increasing the probability of these sequences being bound by transcription factors. Such a model is consistent with data showing that transcriptional activators have the highest affinity for sequences at the "edges" of the nucleosome (Figure 7.2). Moving towards the central regions of the DNA associated with the nucleosome core, the binding of transcription factors becomes increasingly inhibited.

A third major determinant of the affinity of transcription factor binding to nucleosome-bound DNA is the number of transcription factor recognition elements positioned within the DNA on a single nucleosome. Studies using mononucleosomes containing five tandemly repeated transcription factor binding sites have shown that

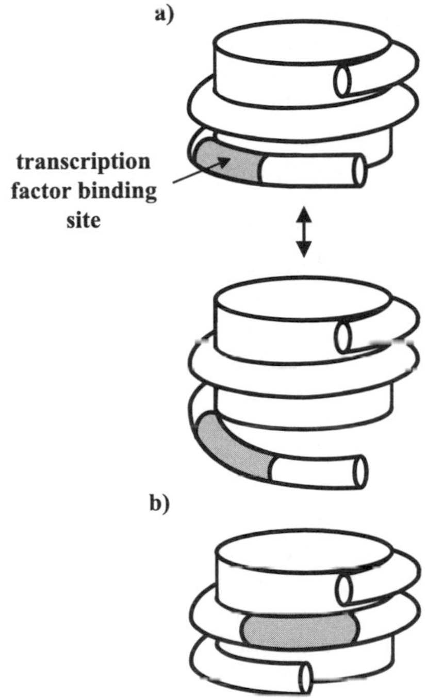

Figure 7.2 Transcription factor access to nucleosome cores.
a) Access for transcription factors to the DNA entering or leaving the nucleosome octamer is aided by the continual movement of the DNA relative to the nucleosome core.
b) However, regions of the DNA centrally placed on the nucleosome core are tightly associated with the protein core, making transcription access much more difficult.

initial binding occurs at the site closest to the edge of the DNA associated with the nucleosome core, followed by binding to the more centrally located sites. The binding to the internally positioned sites within the nucleosome occurs at lower transcription factor concentrations than that required for the binding to mononucleosome templates harbouring single sites in the same internal locations.

Thus, the initial binding of the transcription factor to the edge of the nucleosome appears to facilitate further binding at sites with apparently poorer access. The initial binding of the transcription factor to the DNA on the nucleosome core can enhance the binding of a second protein to the DNA, leading to a co-operative or synergistic effect (Figure 7.3). Such co-operativity may be a general property governing transcription factor–nucleosome interactions.

Presumably, such co-operativity arises from successive loosening of DNA–histone interactions, the first protein bound loosening the outer DNA–histone interactions, allowing a second protein access to its binding site further loosening DNA–histone interactions, and so on.

This leads to a cascade of factor binding to the nucleosomal DNA even at the most central binding sites. In support of this mechanism, the most pronounced co-operative effects appear to be mediated by the binding of activators in close proximity to each other.

2.3 Protein modification

Histone proteins can be readily acetylated and deacetylated, perhaps as part of a regulatory mechanism.

A second, though not necessarily distinct, observation is that nucleosomes can readily be acetylated and deacetylated. Histone acetylation occurs at specific lysine residues and may result in conformational changes in the nucleosome, which in turn might allow access for or facilitate binding of transcription factors. Should the altered nucleosome conformation have a greater affinity for transcription factors, a co-operative effect would then be observed for the binding of additional factors.

The histone core can be divided into two domains; a histone fold domain, involved in histone–histone interactions and in wrapping DNA on nucleosomes, and the amino-terminal tails that lie on the outside of the nucleosome, where they can interact with regulatory proteins.

These amino-terminal tails are rich in lysine residues and are targets for acetylation (chemical changes shown in Figure 7.4). Acetylation reduces the affinity of the histone H4 tail for DNA. Such acetylation will cause a reduction in the wrapping of DNA around the histone octamer and the nucleosome cores will pack together less efficiently in arrays.

For example, the transcription factor TFIIIA does not bind efficiently to a 5S rRNA gene located on a nucleosome if the core histones are not acetylated. Upon acetylation, however, binding is greatly increased.

Histone acetylation seems to provide a molecular mechanism by which DNA binding by transcription factors can be facilitated while still maintaining the basic nucleosomal architecture. Further evidence linking acetylation and transcription factor binding comes from the observation that TFIID has histone acetyltransferase activity.

TFIID is part of the transcription complex; the 250 kDa RNA polymerase II TATA-binding protein (TBP)-associated factor ($TAF_{II}250$) is the core subunit of TFIID and interacts with a variety of other TAFs as well as TBP. $TAF_{II}250$ is necessary for the activation of particular genes and associates with components of the basal transcription machinery, including TFIIA, TFIIE, and TFIIF. In addition, $TAF_{II}250$ has both kinase and histone acetyltransferase functions.

That histone acetylation increases the binding of transcription factors to DNA on a nucleosome core and some transcriptional activators can acetylate histones leads to a model for transcriptional regulation in which recruitment of co-activators could direct the local destabilisation of repressive histone–DNA interactions. Nucleosomal positioning at a gene promoter could affect the gene's transcriptional ability, preventing mRNA production. Targeted acetylation could

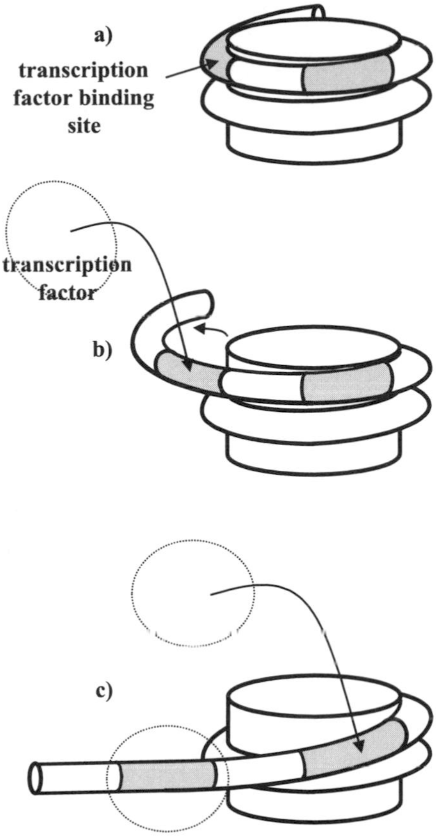

Figure 7.3 Multiple transcription factor binding.

a) The binding of a transcription factor to the outer regions of the DNA associated with the nucleosome core will partially separate the DNA from the core.
b) This will ease access for other transcription factors to bind to DNA more centrally located on the core.
c) Such co-operativity in transcription factor binding allows access to DNA, which would otherwise be inaccessible.

provide a mechanism to allow the basal transcriptional machinery to displace nucleosomes and allow transcription.

It is also known that transcriptional regulators exist that have the ability to *deacetylate* histones. This finding provides a molecular mechanism whereby transcription can be continually controlled. It seems that repressor complexes deacetylate histones and stabilise nucleosomes, whereas activator complexes acetylate histones and disrupt nucleosomes.

Some evidence exists, which though controversial, suggests that core histone proteins remain associated with DNA in the promoter regions even in the presence of the polymerase machinery. Perhaps, then, nucleosomes can be acetylated to allow transcription and deacetylated when the polymerase has passed, giving continuous control.

3. What happens to the nucleosome during transcription?

The question of how transcription factors access their recognition sites within nucleosomal DNA has been explored using genomic

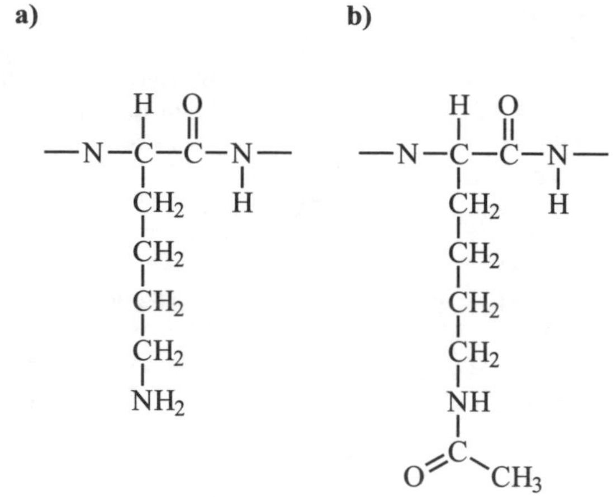

Figure 7.4 Structures of (a) lysine and (b) acetylated lysine.

footprinting data and chromatin reconstitution experiments. Additionally, in recent years, three dimensional structures of the core histone octamer, a histone homodimer, two core histone-like subunits of TFIID, a linker histone, and a linker histone-like transcriptional activator have been determined.

To fully understand the mechanism of transcription, the fate of the nucleosome core during transcription needs to be confirmed. If co-occupancy of nucleosome cores and transcription factors can occur, it may well be that transcription occurs in the presence of histones. *In vivo* studies of the mouse mammary tumour virus long terminal repeat suggests that such a complex of transcriptional activators and histones may exist in an activated state at this gene's enhancer.

Alternatively, the binding of multiple transcriptional activators to other enhancers and promoters may result in DNA–protein complexes that are incompatible with the presence of the histone proteins. Also, the retention of histones on DNA may well be inconsistent with the binding and engaging of RNA polymerase (see below). If this is so, there must be a mechanism for removing nucleosomes from the promoter elements either by removing the nucleosome cores from the DNA or by moving the nucleosome cores along the DNA.

It has been proposed that transcription factors be classified according to their ability to bind to a DNA sequence on a nucleosome. "Initiator" factors are those that can bind their target sequences within nucleosomal structures and initiate chromatin remodelling and transactivation. "Effector" factors are those that are unable to bind their sequence in nucleosomal DNA and depend on initiator factors to alter nucleosomal structure.

The difference between initiator and effector factors seems to be the number of protein–DNA contacts. Initiator factors have only a few DNA contacts clustered on one side of the double helix, whereas effector factors have extensive contacts distributed throughout the whole circumference of the DNA helix.

3.1 Nucleosome sliding?

There are several views as to how the transcribing RNA polymerase deals with nucleosomes.

The simplest mechanism of clearing nucleosomes from the promoter and enhancer regions would be nucleosome "sliding", in which nucleosomes displace from transcription recognition sites in response to factor binding. Some studies have indicated that nucleosomes can reposition themselves over short distances (tens of base pairs). Such repositioning over longer distances is, of course, restricted by the presence of adjacent nucleosomes. Thus, for nucleosome sliding to be truly effective, it would require co-ordinated movement of a large number of nucleosomes to clear the promoter region. There is some evidence for this; it has been shown that nucleosomes can be cleared from DNA in an ATP-driven process. This raises the concept that chromatin is not a stagnant structure as had once been thought, but a dynamic structure in which ATP hydrolysis is used to facilitate transcription factor access.

3.2 Histone dissociation?

"Initiator" factors are those that can bind their target sequences within nucleosomal structures, whereas "effector" factors are unable to bind their sequence in nucleosomal DNA and depend on initiator factors to alter nucleosomal structure first.

A second method of generating a nucleosome-free region to enable transcription factor binding would be simple dissociation of nucleosomes from DNA. While transcription factor binding to nucleosomal DNA does not directly result in nucleosome displacement, factor binding can allow the movement of nucleosomes onto other DNA regions (either *cis* or *trans*).

For example, *in vitro*, at least, nucleosomes can "transfer" to other DNA. The binding of five dimers of a GAL4 derivative to mononucleosome templates permits transfer of the histones onto another DNA fragment, whereas in the absence of factor binding, no such histone transfer occurs. Therefore, histones will not simply leave DNA, nor can they be "forced" off, unless there is another DNA fragment for them to bind to.

There are two potential pathways for histone dissociation, which are dependent on the histone acceptor. In the first, the nucleosome is removed to a *trans* position from a transcription factor binding site. Histone displacement is mediated by a group of acidic proteins, termed nucleosome assembly factors, that bind and transfer histones from and to DNA during nucleosome reconstitution under physiological conditions.

Nucleosome assembly is a stepwise process in which a tetramer of histones H3 and H4 forms on the DNA, followed by the non-covalent association of two heterodimers of H2A and H2B completing the histone octamer (Figure 7.5). *In vitro* studies with two different nucleosome assembly proteins, nucleoplasmin and nucleosome assembly protein 1 (NAP-1), demonstrate that these proteins can facilitate transcription factor binding and nucleosome disassembly.

Nucleoplasmin and NAP-1 enhance transcription factor binding to nucleosome cores by allowing dissociation of one or both H2A/H2B dimers. Also, when multiple transcription factor sites are present on a nucleosome within an array of nucleosomes, nucleoplasmin facilitates disassembly of the nucleosome core with the transcription factor binding site. The H2A/H2B dimers are removed first, followed by the H3/H4 octamer. Thus, the same proteins that donate histones to DNA during nucleosome assembly may participate in chromatin remodelling by disassembling nucleosomes bound by transcription factors.

A second possible pathway for exposing DNA involves the movement of the intact histone octamer, without any disassembly and reassembly. There is some experimental evidence supporting this idea; data suggest that a histone octamer can relocate to the rear of the advancing polymerase, via an intermediate state where the octamer is simultaneously associated with DNA both in front of, and behind, the polymerase.

It has been shown that the position of nucleosome cores was changed after transcription, supporting the displacement pathway. However, it remains to be determined whether there is a bridged intermediate between the core DNA sites or whether the nucleosome core moves off the DNA and then rebinds after the polymerase has passed.

A histone octamer can step around a transcribing polymerase over a distance of 40–95 base pairs, without leaving the DNA template. The mechanism proposed involves a bridging complex in which the octamer is in contact with both donor and acceptor DNA sites, leaving the transcribing polymerase in between the two sets of histone-DNA contacts. This bridging complex would be energetically favoured relative to a free octamer intermediate, because histone-DNA contacts would be broken sequentially rather than simultaneously. The model also involves a spooling mechanism for octamer translocation during transcription. As the transcribing polymerase approaches the nucleosome core, the DNA begins to uncoil from the histone protein surface. As the polymerase continues to move on to the nucleosome core, further DNA will uncoil and become captured behind the polymerase on the exposed histone octamer surface, resulting in a DNA loop within the nucleosome core. The polymerase is bound within this internal loop of between 40 and 95 base pairs. The DNA ahead of the transcribing polymerase continues to uncoil from the octamer, while the DNA behind coils back around the protein when the polymerase has passed. Thus, the nucleosome core is translated to its new position and the polymerase is able to complete the transcript (mechanism is outlined in Figure 7.6).

3.3 Specialised nucleosomes

In addition to histones H2A, H2B H3, and H4, which comprise the typical nucleosome, other histone-like proteins are known. The histone proteins H2A, H2B, H3, and H4 apparently evolved from DNA-binding proteins that contained only the so-called "histone fold" domain and lacked any form of extended amino tail. In fact, the archaebacteria contain a DNA-binding protein, which contains *only* the histone-fold section; this protein has the ability to wrap DNA around itself, forming a nucleosome-like structure. The histone proteins H2A, H2B, H3, and H4 have retained this ability, but have also taken on a second role, allowing them to interact with other proteins.

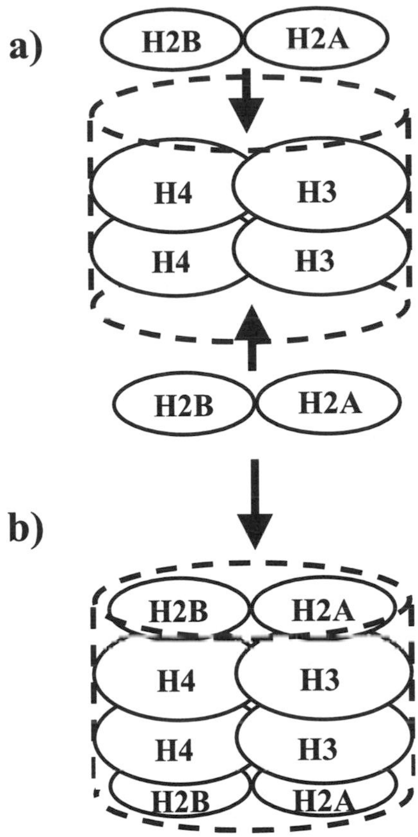

Figure 7.5 Nucleosome assembly.

a) The histone octamer is assembled in stages; two molecules each of histones H3 and H4 associate; DNA then associates with this tetramer.

b) Two heterodimers of H2A and H2B then bind, completing the histone octamer.

Other specific regulatory proteins have made use of their histone-fold domains to confer very specialised properties on individual nucleosomes through the replacement of normal histones within chromatin. These regulatory proteins contain the histone-fold domain and make use of it both to direct specific protein–protein interactions and to bind to DNA.

A number of the TATA-binding protein-associated factors (TAFs) contain a similar histone-fold structure, for example, $TAF_{II}40$ and $TAF_{II}60$. $TAF_{II}40$ and $TAF_{II}60$ exist as heterodimers in TFIID. $TAF_{II}40$ resembles H3 and $TAF_{II}60$ resembles H4.

Additionally, both $TAF_{II}40$ and $TAF_{II}60$ have an extended carboxyl-terminal domain; this interacts with other subunits of TFIID as well as transcriptional activators. In fact, it has been proposed that $TAF_{II}40$ and $TAF_{II}60$ are involved in forming a nucleosome-like octameric structure, which excludes normal nucleosomes from the TATA box region. This structure maintains the DNA in a semi-compacted state, which facilitates transcription.

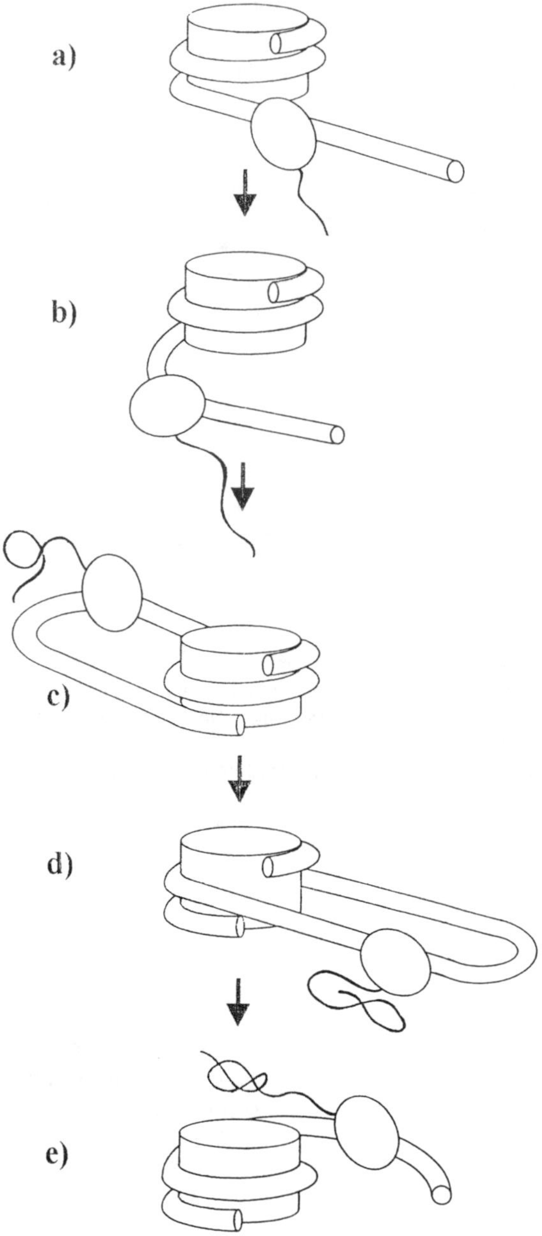

Figure 7.6 A potential mechanism for transcription through nucleosome cores. Modified from Studitsky *et al.* (1994).

a) RNA polymerase binds and initiates transcription from a gene promoter.
b) As the polymerase reaches a nucleosome it induces the dissociation of DNA from the core in a stepwise manner. This mechanism may be initiated by RNA polymerase or by transcription factors.
c) The DNA behind the polymerase binds to the exposed octamer surface to form a loop within the nucleosome core to which the transcribing polymerase is still bound.
d) The DNA ahead of the transcribing polymerase continues to uncoil from the octamer and the DNA behind the polymerase begins to coil around the octamer.
e) The nucleosome core is re-formed behind the polymerase, which has been translated 40 to 95 base pairs.

Summary

When the three eukaryotic RNA polymerases bind DNA, they are presented with an apparent obstacle course. Eukaryotic DNA is already bound by many proteins that co-ordinate the higher order structures of DNA that enable 2 metres of DNA to be efficiently packaged in a cell measuring a few micrometres. Such proteins may compete with the polymerase or with transcription factors for DNA sequences and/or may simply physically (sterically) block access to a promoter. Despite this, it seems that the RNA polymerase can transcribe right through nucleosomal DNA. It is not entirely clear what happens to the nucleosome during this process, but it is effectively moved out of the way of the polymerase. However, it should not be thought that the histones are mere scaffolding; they play a role in regulating transcription, as is evident from studies in yeast with mutant histones.

FURTHER READING

1. Grunstein, M. (1997) Histone acetylation in chromatin structure and transcription. *Nature* **389**, 349–352.
2. Perez-Martin, J. and De Lorenzo, V. (1997) Clues and consequences of DNA bending in transcription. *Annu. Rev. Microbiol.* **51**, 593–628.
3. Beato, M. and Eisfeld, K. (1997) Transcription factor access to chromatin. *Nucleic Acids Res.* **25**, 3559–3563.
4. Werner, M. H. and Burley, S. K. (1997) Architectural transcription factors: proteins that remodel DNA. *Cell* **88**, 733–736.
5. Edmondson, D. G. and Roth, S. Y. (1996) Chromatin and transcription. *FASEB J.* **10**, 1173–1182.
6. Suzuki, M., Loakes, D. and Yagi, N. (1996) DNA conformation and its changes upon binding transcription factors. *Adv. Biophys.* **32**, 53–72.
7. Rippe, K., Von Hippel, P. H. and Langowski, J. (1995) Action at a distance: DNA-looping and initiation of transcription. *Trends Biochem. Sci.* **20**, 500–506.
8. Owen-Hughes, T. and Workman, J. L. (1995) Experimental analysis of chromatin function in transcription control. *Crit. Rev. Eukaryotic Gene Expr.* **4**, 403–441.
9. Grunstein, M. (1990) Nucleosomes: Regulators of transcription. *Trends Genet.* **6**, 395–400.
10. Li, Q., Wrange, O. and Eriksson, P. (1997) The role of chromatin in transcriptional regulation. *Int. J. Biochem. Cell Biol.* **29**, 731–742.
11. Studitsky, V. M., Clark, D. J. and Felsenfled, G. (1994) A histone octamer can step around a transcribing polymerase without leaving the template. *Cell* **76**, 371–382.

END OF UNIT QUESTIONS

1. Describe the structure of the nucleosome.
2. What happens to the nucleosome during transcription?
3. Discuss the role of $TAF_{II}250$ in eukaryotic transcription.

Problems

1. Discuss the role of chromatin in transcription (see references 8 and 10).
2. Discuss chromatin remodelling (see reference 4).

8 Bacteriophage Lambda (λ): A Transcriptional Switch

OBJECTIVES

a) describe the transcriptional control of the life cycle of bacteriophage λ

1. Overview

Control of the life cycle of the bacteriophage lambda (λ) is mediated through a transcriptional switch. Phage λ is important as a model system that has been completely characterised. The elements of the system and its mechanism of operation will be discussed in this chapter.

2. Phage λ

Phage λ is a virus that infects certain bacteria.
A transcriptional switch controls the two parts of its life cycle.

Bacteriophage λ is a specialised virus that infects certain bacteria. It has been heavily studied for many years as a model system and much is known about its genome and life cycle. The genes of phage λ are contained within a single DNA chromosome itself contained within the protein coat of the phage. The λ genome is a linear double-stranded DNA of 48.5 kilobases, with 12-base complementary single-stranded sticky ends (the cohesive (cos) sites).

The protein coat is an elaborate structure with a "head" and a "tail", comprising at least 15 different proteins, encoded within the single chromosome. The phage coat and tail do not enter the cell. Phage λ enters the *E. coli* cell to undergo replication. It is therefore an obligate parasite.

The phage particle infects *E. coli* cells by attaching to a protein on the cell surface via its tail, and "injecting" its chromosome into the bacterium (Figure 8.1). Specifically, λ interacts with the *E. coli* malB protein, which ordinarily transports maltose into the bacterium when it is available in the growth medium. As a consequence, some bacteria carrying defects in the malB protein–or defects in the gene leading to no expression of it–are immune to infection by phage λ.

2.1 The phage life cycle

In the lytic stage, phage λ DNA is replicated, its genes for the coat proteins are expressed, mature phage assembles in the host bacterium, and the bacterium is then lysed releasing the phage particles.

After infection by the phage λ chromosome, the bacterial cell can be made to follow one of two alternative pathways. These are the so-called lytic and lysogenic pathways. Once the λ genome is inside the bacterium, the DNA circularises, using the 12-base cos ends. The bacterium's DNA ligase then seals the nicks and the DNA becomes supercoiled.

In certain circumstances the phage will enter its lytic life cycle. The lytic process activates various sets of phage genes allowing the phage λ chromosome to be replicated and new head and tail proteins to be synthesised. From these components, the formation of new phage particles occurs. This process can be completed within approximately 45 minutes of the initial infection. At this point the bacterium will be lysed, releasing about 100 progeny phage per cell, each of which are capable of infecting a further bacterium.

The second possible pathway for the phage is to enter the lysogenic life cycle. In the lysogenic pathway all but one of the phage genes are switched off, and the phage chromosome, now referred to as a

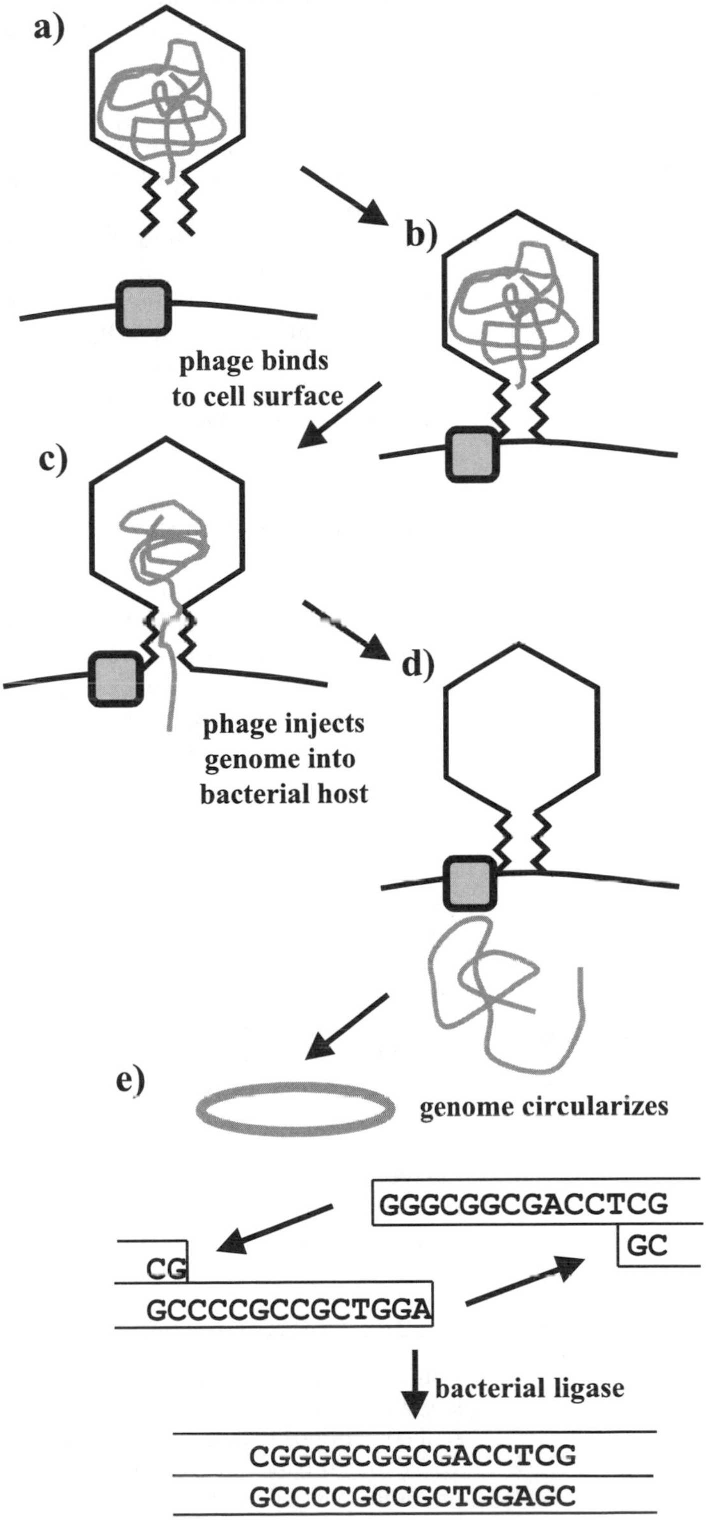

Figure 8.1 Phage λ infection.

a) Phage λ interacts with the bacterial outer membrane.
b) The phage recognises the bacterial malB protein and binds to the outer membrane.
c) The phage injects its genomic DNA into the bacterium.
d) The "head" and "tail" of the phage do not enter the cell.
e) Following injection of the phage genome into the bacterium, the phage DNA circularises, using the 12 base cos (cohesive) ends. The bacterial DNA ligase then seals the nicks.

In the lysogenic stage, the phage DNA is integrated into the bacterial chromosome and is duplicated and passed on to all progeny bacteria.

prophage, integrates into the host bacterial chromosome. This additional sequence residing in the bacterial genome will have no or very little effect on the bacteria, which will continue to grow and divide as before. However, in such bacterial replication, the prophage will be passively replicated and distributed to all progeny bacteria. This process can continue indefinitely; if the bacterial environment remains constant, such growing bacteria only rarely produce phage.

However, exposure of these bacteria to ultraviolet (UV) radiation or other DNA-damaging stimuli will cause the prophage to change from the lysogenic life cycle to the lytic cycle and produce a new crop of phage. The induction is caused by the lysogen sensing that the host cell has been damaged and initiating the lytic pathway in response. This chapter will illustrate the mechanisms of this switching from the lysogenic to the lytic life cycle and the control of the switch.

3. Overview of the genetic switch

In the lysogenic life cycle only one gene from the phage λ chromosome is active; the protein encoded by this gene is known as the λ repressor. The repressor acts as both a positive and a negative regulator of gene expression. By binding to just two operators on the λ chromosome it turns off all other phage genes while it simultaneously turns on expression of its own gene.

In the prophage cell, there is a single copy of the repressor gene; from this are produced approximately 100 molecules of the λ repressor protein. This large excess is present so as to bind any additional λ chromosomes that might enter the cell, effectively preventing secondary infection by further phage.

The "decision" as to which pathway to follow – lytic or lysogenic – is controlled by two DNA-binding proteins, Cro and the λ repressor.

Irradiation of the lysogenic cells with UV light inactivates the repressor. As a result a second phage regulatory protein, Cro, is synthesised; Cro is required for the lytic life cycle. Cro also binds specific DNA regions, the same ones as does the λ repressor, but Cro binding has the opposite physiological effect.

These two regulatory proteins, together with RNA polymerase and their promoter and operator sequences in the λ chromosome, constitute the "genetic switch". The switch has two positions; in the first—the lysogenic state–the λ repressor is produced and synthesis of Cro protein is inhibited, whereas in the second, Cro is expressed and the synthesis of the λ repressor is inhibited.

3.1 Cro and λ repressor operators and promoters

Within the λ chromosome the genes encoding the λ repressor and Cro, known as *cI* and *cro*, respectively, are adjacent. The two genes are transcribed in opposite directions and do not overlap; in fact, they are separated by 80 base pairs. Within this 80-base-pair region are located promoters and operators for each gene, including the binding sites for both the λ repressor and the Cro proteins.

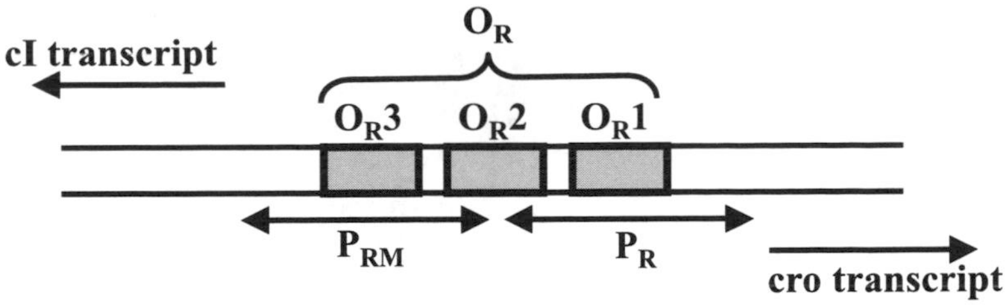

Figure 8.2 Operator structure. The operator contains three distinct regions, O_R1, O_R2, and O_R3, to each of which λ repressor and cro bind.

The two genes have their own promoters; that for *cI* is known as P_{RM} and the *cro* promoter is known as P_R. The two promoters are adjacent, but not overlapping.

3.1.1 *The operator*

The operator, O_R, spans both the Cro and repressor promoters. The operator contains three similar regions, O_R1, O_R2, and O_R3 (Figure 8.2). The λ repressor and cro both bind to these regions to regulate the activities from the promoters P_{RM} and P_R.

The "operator" is a key DNA sequence that overlaps both the Cro and λ repressor promoters. The Cro and λ repressor proteins bind to three sites within the operator.

The individual operator sites are 17 base pairs in length; their sequences are similar but not identical, and the λ repressor and cro proteins can distinguish between them. The two proteins have slightly different affinities for the sites; as a result, at different concentrations of repressor or cro, the two proteins will occupy different operator sites.

The RNA polymerase used to transcribe these, and other phage genes, is the bacterial enzyme; the phage does not have its own. If the P_R promoter is in its active state and RNA polymerase transcribes, cro mRNA will be produced; if the polymerase binds to P_{RM} in its active state, repressor will be produced. However, the two promoters can never be occupied simultaneously. The polymerase can only bind to one or the other, depending on the position of the switch; that is, λ repressor and cro are never transcribed at the same time.

There is an important difference between the two promoters. At P_R transcription will occur without the need for activator proteins. This is not so at P_{RM}; there, an activator is needed, a role filled by the λ repressor protein.

3.2 Repressor and cro structures

Before discussing the mechanism of action of the switch it is necessary to first outline the gross structural characteristics of the cro and repressor proteins. Cro comprises 66 amino acids, and folds into a single domain. Cro exists as a homodimer and binds to λ DNA in the same position as the λ repressor. Only cro *or* the repressor can bind to a particular site, never both. These two proteins oppose each other and play opposite roles in the switch mechanism. The λ repressor protein is 236 amino acids in size and folds into a structure resembling a dumbbell, with two domains separated by a string of 40 amino acids (Figure 8.3a).

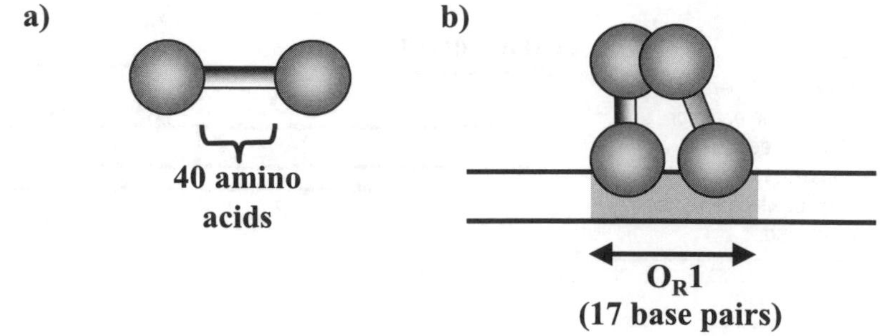

Figure 8.3 Structure of the λ repressor.
a) the "dumbbell" appearance of repressor protein.
b) repressor homodimer binding to an operator sequence.

These two domains correspond to the amino- and carboxyl-termini of the protein. The repressor also exists primarily as a homodimer; the carboxyl-terminal domains interact strongly. The amino-terminal regions also interact with each other; however, their primary role is in contacting DNA (Figure 8.3b). One dimer can bind to each of the three sequences in O_R.

3.3 Co-operativity of repressor binding

λ repressor dimers bind co-operatively to the O_R sites within the operator. The binding of the first repressor to O_R1 causes a conformational change in the DNA such that the second λ repressor binds to O_R2 with higher affinity.

The λ repressor is capable of binding all three operator regions within the promoters; however, the affinity of repressor for each site is different. Specifically, the repressor has the highest affinity for O_R1, approximately ten times higher than that for O_R2 or O_R3. For activation of the P_{RM} promoter and hence production of more repressor, the repressor needs to bind to both O_R1 and O_R2. Logically then, there is a problem; if there is a ten-fold difference in affinity between O_R1 and O_R2, there would need to be a ten-fold increase in concentration of repressor to allow its own production. However, as the P_{RM} promoter is inactive there can be little increase in repressor concentration.

The phage overcomes this potentially fatal problem by using interactions between repressor dimers; when the first repressor binds to O_R1 (Figure 8.4a), it immediately causes a change in the DNA structure such that further λ repressor binds to O_R2 with higher affinity. The second λ repressor dimer not only binds O_R2, but also interacts with the repressor dimer bound to O_R1 (Figure 8.4b). As a result of this interaction, the O_R1 and O_R2 sites fill almost simultaneously despite the *apparent* ten-fold difference in affinity.

The binding of the second λ repressor activates P_{RM}, allowing transcription of the mRNA for the repressor leading to an increase in concentration of the repressor protein. When the concentration increases considerably, even O_R3 will become occupied by repressor protein (Figure 8.4c). This will have the *opposite* effect to repressor binding at O_R2 and will *inactivate* the P_{RM} promoter, halting production of further repressor. The λ repressor binds to site O_R3 more weakly than to O_R2 *in vivo*, even though the two sites have virtually equal intrinsic affinities for the repressor. The reason for this is that the binding to O_R2 does not increase binding affinity to O_R3, as is seen between O_R1 and O_R2. Repressor binding to O_R3 is essentially *independent* of binding to sites O_R1 and O_R2. The reason for the lack of co-operativity between repressors at

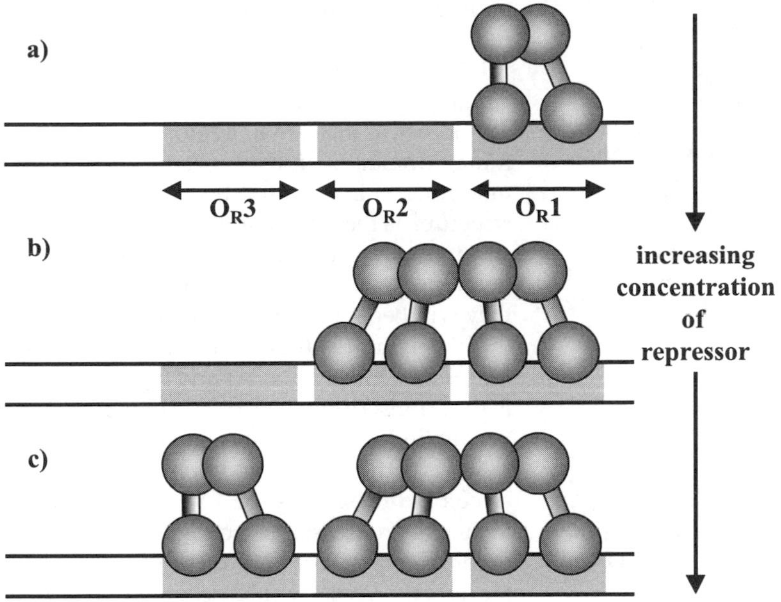

Figure 8.4 λ repressor binding.
a) The first λ repressor dimer binds to O_R1.
b) Co-operative binding of the second λ repressor dimer activates promoter P_{RM}.
c) Repressor binding to O_R3 causes inactivation of the P_{RM} stopping repressor production.

O_R2 and O_R3 lies in the nature of the interactions between adjacent dimers. Once a repressor dimer at O_R2 interacts with the dimer at O_R1, it is unable to [also] interact with a dimer on O_R3.

3.4 Lysogeny or lysis?

In the lysogenic phase, λ repressor will occupy sites O_R1 and O_R2, allowing continuous production of repressor and inhibiting the expression of Cro by blocking the promoter P_R.

In the lysogenic life cycle of phage λ, O_R1 and O_R2 will be occupied by the phage repressor, which will continuously allow the production of repressor and also inhibit the production of Cro by blocking transcription from the promoter P_R. The repressor will be constantly synthesised as the cells grow and divide, whereas the *cro* gene will be continually inhibited. If the cell's growth slows, the repressor concentration will increase and the repressor will begin to bind to O_R3 as well as O_R1 and O_R2. Binding to O_R3 inhibits transcription from the *cI* gene and therefore the production of repressor. As the rate of cell division increases, the concentration of the repressor will drop and the repressor will no longer bind to O_R3, and this will allow repressor synthesis to begin again.

The repressor dimers bound to operator sites are not static; they will leave and either rebind or be replaced by another repressor molecule. This is a process that will occur constantly. However, the repressor concentration within a wild-type cell is high enough so that O_R1 and O_R2 will be continuously filled and allow production of repressor. This balance can be disturbed by outside influences, and the switch from lysogeny to lysis will be made.

Lysis can be induced by ultraviolet light. Ultraviolet (UV) light will damage DNA; such damage leads to changes in the function of a protein known as RecA. Under normal cellular conditions RecA

Lysis is induced typically by DNA damage. Such damage leads to a change in the role of the RecA protein. When DNA is damaged, RecA becomes a protease that cleaves the repressor so that it is no longer functional.

is used by the bacterium to catalyse recombination between DNA molecules.

If DNA is damaged, the RecA protein changes function and becomes a highly specific protease that cleaves repressor monomers. The cleavage occurs in the "stem" region between the carboxyl- and amino-terminal domains of the protein. When this separation occurs the monomer is inactivated because it is now unable to form an active structure as the interaction between the amino-terminal domains is not strong enough to form stable dimers without the carboxyl-terminal domains. As the concentration of the active repressor begins to drop, they will begin to leave the operator and not be replaced.

This action results in two changes, the first being the stopping of repressor synthesis as O_R1 and O_R2 are no longer occupied. The second is that transcription from the *cro* gene is no longer inhibited and Cro production will begin, moving the host cell towards lysis.

Cro acts by binding to the same sites with the operator that the λ repressor binds to, but leads to lysis.

Cro is active as a homodimer. Its actions are transmitted through the same binding sites as the λ repressor, namely the three operator sites. Unlike the repressor, its actions are strictly as a negative regulator. The order of affinity for Cro to the operators is reversed compared to the repressor, *i.e.*, O_R3 will be occupied first. Also, for Cro binding to these three sites there is no co-operativity; binding to each operator site is independent of the others. Cro binding to O_R3 prevents any binding of the RNA polymerase to the P_{RM} promoter and therefore production of repressor. The Cro protein's production begins a process that cannot be stopped or reversed. When Cro has been produced it is inevitable that the lytic pathway will be followed.

Cro is used by phage λ in the early stages of the lytic life cycle. At high concentrations, Cro has the ability to control its own synthesis. As the concentration of Cro increases, it will bind to operators O_R1 and O_R2. When this occurs, synthesis of Cro will be inhibited. Cro, therefore, first turns off λ repressor synthesis and then, later, down-regulates expression of itself.

4. How does the phage decide?

So far we have described the mechanism of the switch, from the lysogenic to the lytic pathway. A further question that must be asked is: when the bacteriophage enters the bacterium, how does it know whether to move into the lytic or the lysogenic life cycle? There must be a control mechanism, which allows the virus to decide.

For a phage to initially grow lytically it must replicate, package its DNA into phage particles, and lyse the bacterium. It must also inhibit the synthesis of λ repressor. For lysogeny to occur the phage must synthesise the enzymes that integrate its chromosome into that of the host, begin synthesis of repressor, and prevent expression of the genes needed for lytic growth.

The initial stages of lytic and lysogenic life cycles are identical, but at a critical step the state of the host will be determined and the events following this decision process will lead to either the lytic or lysogenic life cycle. Each of these pathways involves the activation of gene cascades, where a regulatory protein typically turns on or off a block of genes. The product from one of these genes will again act as a regulatory protein for further gene control.

4.1 Infection of the bacterium

The first stage of either the lytic or the lysogenic life cycle is the injection of the λ chromosome into the host cell. The chromosome will immediately circularise, using cohesive ends present at either end of the chromosome. These strands are then ligated using a bacterial enzyme. The cohesive ends consist of 12 base pairs shown in Figure 8.1.

4.2 Transcriptional control

Gene expression within phage λ can be divided into three stages: "very early", "early", and "late" gene expression. In the very early stage of gene expression only the genes *N* (N-positive regulator) and *cro* are transcribed. As the phage moves into the early stage, the number of active genes increases to include the genes needed for DNA recombination and replication.

In lysogeny, the λ chromosome does not insert randomly in the *E. coli* genome, but at a specific point.

As the phage moves into this phase of growth the phage can follow two alternative pathways, lytic or lysogenic. If the phage chooses the lytic pathway the early genes are turned off and the head, tail, and lysis genes become active. New phage particles are formed and are released when the bacterium is lysed. If the phage chooses the alternate lysogenic pathway, then only two genes will be active, *cI* (repressor) and *int* (integrase). The *int* gene product will allow integration of the phage chromosome into the host. After this has occurred only *cI* will be active.

Chromosome integration is not a random event and is controlled by the enzyme "integrase". Integration occurs only at specific positions within both the phage and bacterial chromosomes. The site within phage is known as *attP* (for phage attachment site) and the corresponding site within the bacteria is known as *attB*. Upon switching to a lytic life cycle this process will be reversed and the circular λ chromosome will be reformed. The excision of the chromosome again requires Int, and also a second protein called Xis.

4.2.1 *"Very early" genes*

Immediately after phage infection, the host bacterium's RNA polymerase will transcribe the *N* and *cro* genes and the corresponding proteins will be synthesised. The promoters for the *N* and *cro* genes are known as P_L, and P_R respectively (Figure 8.6a). These promoters obtained their names for historical reasons; "L" signifies "left" and R "right" of the repressor gene, *cI*.

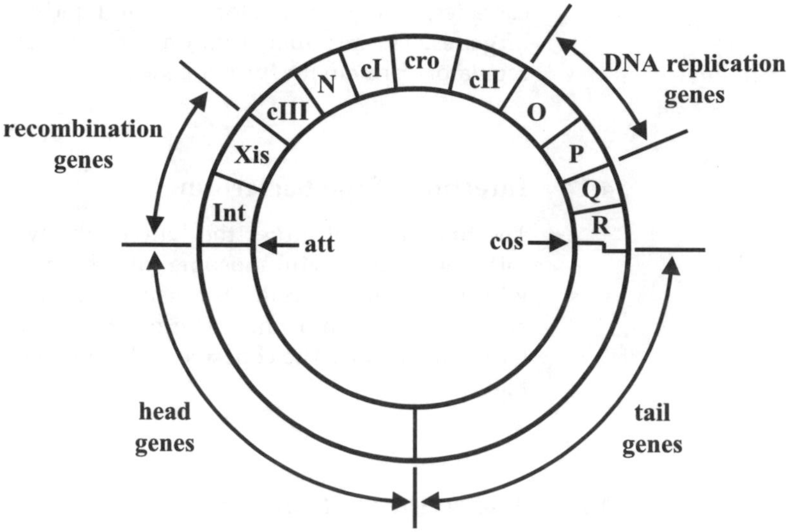

Figure 8.5 Genetic map of phage λ.

4.2.2 *"Early" genes*

The N protein is a positive regulator and it turns on the genes left of *N*, that is *cIII, Xis,* and *int* (*Xis* and *int* are needed for DNA recombination). N also switches on the genes to the right of *cro, cII, O, P,* and

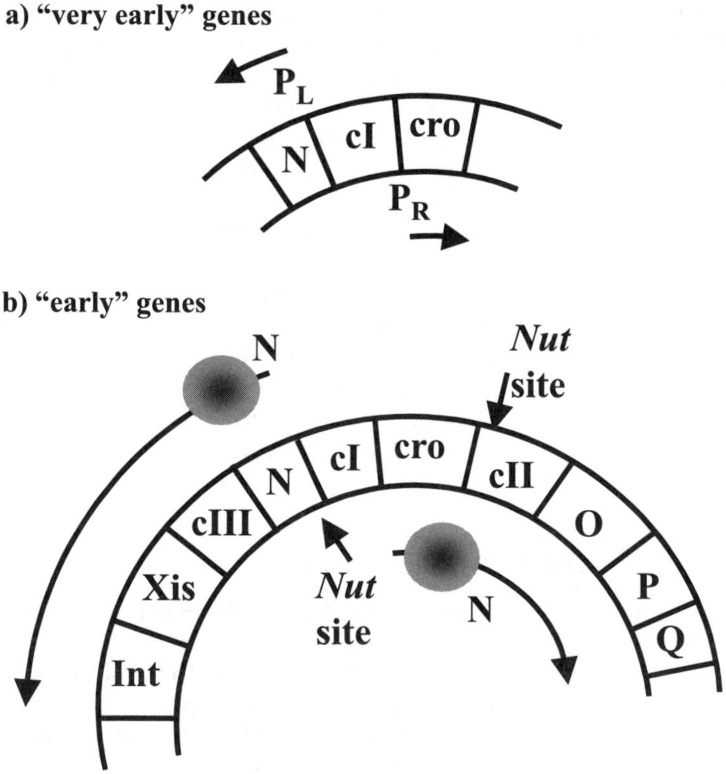

Figure 8.6 Gene expression occurs in phases.
a) "Very early" genes. Immediately after infection, the host RNA polymerase transcribes the *N* and *Cro* genes from the promoters P_L and P_R, respectively.
b) "Early" genes. The anti-termination activity of N allows transcription from promoter P_L to continue through N to produce transcripts including genes *cII*, *Xis*, and *Int*. This activity also allows transcription from the promoter P_R to continue beyond *cro* to produce transcripts of the genes *cII*, *O*, *P*, and *Q*.

Q (proteins O and P are involved in DNA replication). The head and tail genes, however, remain inactive. The N protein allows RNA polymerase to transcribe through DNA sequences that were previously RNA polymerase termination sites, *i.e.*, N acts as an anti-terminator protein. In the presence of N, the mRNA produced from the *N* and *cro* genes is extended through the adjacent genes (Figure 8.6b). The extension of the N mRNA allows the transcription of the *cIII*, *Xis*, and *Int* genes and the production of those proteins. Similarly, the extension of the Cro mRNA leads to the formation of the cII, O, and P proteins.

The N protein recognises a specific DNA sequence called *Nut* (*N utilisation*) and this allows the polymerase to read through termination signals. However, this effect is specific and does not allow the polymerase to read through *all* termination signals. There are two *Nut* sites, the first at the beginning of the *N* gene, the second to the right of the *cro* gene. At this point the phage must decide which pathway it is to follow, lytic or lysogenic.

4.2.3 *"Late" genes; the lytic life cycle*

The production of the Q protein in "early" gene expression allows expression of the late lytic genes (Figure 8.7). Like N, Q also works as an anti-terminator protein and allows production of a specific mRNA beginning at the P_R' promoter, located to the right of the *Q* gene. In the absence of Q, a short mRNA strand is transcribed, but in its presence the mRNA is extended through the head and tail protein genes, allowing the formation of the phage coat proteins.

While the late lytic genes are being transcribed and their proteins are being translated, Cro protein binds to O_R3 (described above) and stops production of the λ repressor. Cro also binds to a second operater, O_L; such binding stops transcription from P_L, blocking the production of cIII, xis, and int. As the concentration of Cro increases, it fills the O_R2

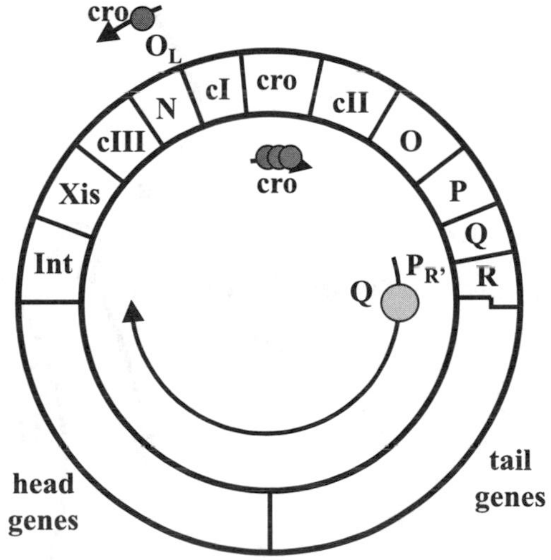

Figure 8.7 "Late" genes (lytic pathway). Production of Q protein in "early" gene expression allows expression of the late genes from promoter P_R. Without Q, only a short transcript is produced. Simultaneously, the concentration of cro increases, allowing binding to O_R1 and O_R2, thereby blocking transcription of its own gene and *cII*, *O*, *P*, and *Q*. Cro also binds to the operator O_L, blocking transcription of the *N* gene and consequently transcription of *cIII*, *Xis*, and *Int*.

and O_R1 sites, repressing P_R and inhibiting its own synthesis and that of cII, O, P, and Q. However, by this time, there is already sufficient Q to allow the formation of the head and tail proteins, which wrap up the newly synthesised DNA molecules. The cell is then lysed and daughter phage are released.

4.2.4 *The lysogenic life cycle*

The cII protein, produced from "early" gene expression, can specifically turn on expression from the *cI* and *int* genes, by allowing RNA polymerase to bind and begin transcription at two promoters that would otherwise be silent, P_{RE} (promoter for repressor establishment) and P_I.

While the repressor protein is able to control the level of its own synthesis, it is unable to initiate its own synthesis. This is a result of the fact that the gene *cI* (repressor gene) can be transcribed from two promoters, one of which is activated by the λ repressor (described previously). The second is activated by the cII protein (PRE) (Figure 8.8). Beginning at PRE, the polymerase transcribes leftward to the end of the repressor gene *cI*. When this mRNA is translated it produces repressor, but not cro, as in this situation the *cro* gene would be transcribed "backwards".

The initial production of repressor therefore originates from the PRE promoter. As the concentration of the λ repressor increases, it binds to O_R1 and O_R2, allowing its own production from the P_{RM} promoter.

cII also causes transcription from the promoter P_I; this causes production of the int protein, which allows the phage chromosome to be integrated into the bacterial host's chromosome.

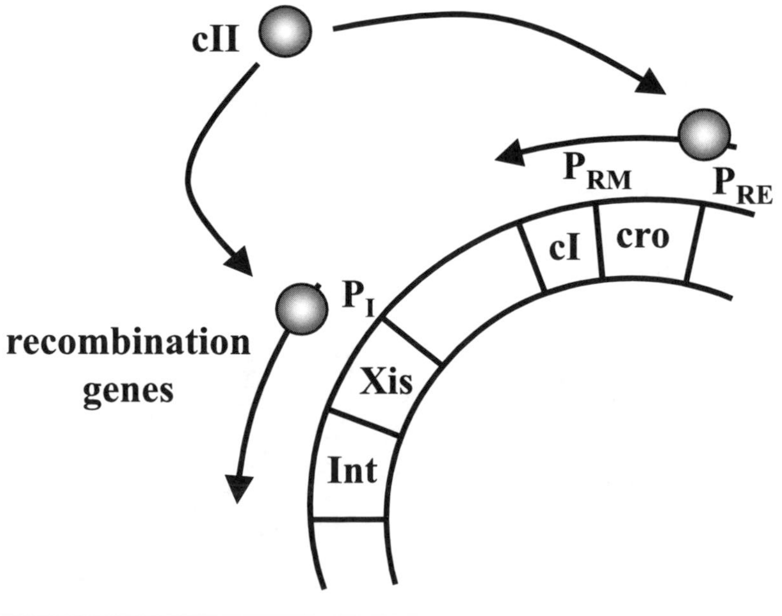

Figure 8.8 "Late" genes (lysogenic pathway). The cII protein can specifically turn on expression from the *cI* and *Int* genes, by allowing RNA polymerase to bind and begin transcription at two promoters that would otherwise be silent, P_{RE} and P_I. This transcription allows the establishment of repressor and the production of the int protein, allowing the phage chromosome to be integrated into the bacterial host's chromosome.

4.3 The decision process

So far we have explained the two possible outcomes of a phage λ; what remains to be explained is how the decision to follow one or the other pathway is made. It seems that the decision is made by the protein cII.

If the production of cII is high after infection of the host cell by the phage then the phage will enter the lysogenic life cycle, whereas if cII activity is low the lytic life cycle will be followed. The cII protein is relatively unstable, and is susceptible to protease digestion. Many proteases are present in the host cell and environmental conditions, such as the availability of nutrients, will affect their concentration. If the host cells are in a nutrient-rich environment, proteases will be active and the phage is more likely to enter the lytic life cycle. If the host cells are growing in a nutrient-poor medium, protease activity will be lower and the phage is likely to enter the lysogenic life cycle. The role of the cIII protein is interesting, as one of its roles is the protection of the cII protein; as a consequence, it also assists in establishing the lysogenic life cycle.

Summary

Phage λ infects bacteria by interacting with a bacterial cell surface protein. Once the phage's DNA is inside the bacterium, it can follow one of two alternative life cycles. In lysogeny, the λ DNA integrates into the bacterial chromosome and is copied to future generations of bacteria every time the bacterial chromosome is duplicated in mitosis. In the lytic pathway, the phage DNA is replicated, coat proteins encoded by the λ genome are expressed and mature phage particles are assembled, before the host bacterium is lysed, releasing new phage into the medium. The decision as to which path to follow is made at the level of transcription, by the binding of cro and the repressor protein to the operator sequence.

Further reading

1. Taylor, K. and Wegrzyn, G. (1995) Replication of coliphage lambda DNA. *FEMS Microbiol. Rev.* **17**, 109–119.
2. Catalano, C. E., Cue, D. and Feiss, M. (1995) Virus DNA packaging: the strategy used by phage lambda. *Mol. Microbiol.* **16**, 1075–1086.
3. Hochschild, A. (1994) Transcriptional activation. How lambda repressor talks to RNA polymerase. *Curr. Biol.* **4**, 440–442.
4. Roberts, J. W. (1993) RNA and protein elements of *E. coli* and lambda transcription anti-termination complexes. *Cell* **72**, 653–655.
5. Friedman, D. I. (1992) Interaction between bacteriophage lambda and its *Escherichia coli* host. *Curr. Opin. Genet. Dev.* **2**, 727–738.
6. Murialdo, H. (1991) Bacteriophage lambda DNA maturation and packaging. *Annu. Rev. Biochem.* **60**, 125–153.

7. Ptashne, M., Jeffrey, A., Johnson, A. D., Maurer, R., Meyer, B. J., Pabo, C. O., Roberts, T. M. and Sauer, R. T. (1980) How the lambda repressor and cro work. *Cell* **19**, 1–11.
8. Sauer, R. T., Jordan, S. R. and Pabo, C. O. (1990) Lambda repressor: a model system for understanding protein-DNA interactions and protein stability. *Adv. Protein Chem.* **40**, 1–61.
9. Weiss, M. A., Karplus, M., Patel, D. J. and Sauer, R. T. (1983) Solution NMR studies of intact lambda repressor. *J. Biomol. Struct. Dyn.* **1**, 151–157.
10. Johnson, A. D., Poteete, A. R., Lauer, G., Sauer, R. T., Ackers, G. K. and Ptashne, M. (1981) λ repressor and cro – components of an efficient molecular switch. *Nature* **294**, 217–223.

END OF UNIT QUESTIONS

1. Discuss the binding of the λ repressor to the operator sequence.
2. How does the phage decide between the lytic and lysogenic pathways?
3. What role does the malB protein play in λ infection?

PROBLEMS

1. Compare and contrast the effects of Cro and the λ repressor (see reference 7).
2. Discuss how the structure of the λ repressor relates to its function (see reference 9).

9 The Lactose Operon in *Escherichia coli*

Objectives

a) introduce a second classic prokaryotic transcription system - the lactose operon in *E. coli*

1. Overview

The lactose operon in *E. coli* is another classic transcriptional system that is now understood in great detail and is a useful model. It illustrates many of the principles described earlier in this book.

2. Bacterial carbon sources

Bacteria can survive on many carbon sources and they often have specialised enzymes to metabolise given carbon sources.

Typically bacteria are exposed to an ever-changing environment in which nutrient availability may change radically. Bacteria respond to such changes in their environment by changing their pattern of gene expression; they express different anabolic and catabolic enzymes depending on what carbon sources and other nutrients are available in the medium.

It would be extremely inefficient and wasteful to synthesise degradative (catabolic) enzymes unless the carbon source for the enzymes is present in the environment. Thus, it would not be efficient to produce lactose-metabolising enzymes unless there was lactose available in the medium and further that there was no or insufficient glucose (the preferred carbon source). However, should lactose be the only carbon source available, the bacterium must quickly induce its lactose-metabolising enzymes or die.

Likewise, it would be extremely inefficient and wasteful to synthesise biosynthetic (anabolic) enzymes if the end product of the biosynthetic pathway was present in the medium and could simply be taken up. Thus, it would not be efficient to synthesise an amino acid such as tryptophan if it was available in the medium. However, should it cease to be available, the bacterium must quickly induce its tryptophan-synthesising enzymes or die.

So, between wastefulness and death, the bacterium has regulated and readily reversible gene expression. Furthermore, this regulation occurs at the level of transcription.

The best known and most studied cases of such regulated expression involve the enzymes of lactose metabolism and tryptophan biosynthesis in the bacterium *E. coli*. The enzymes of lactose metabolism will be discussed here; those of tryptophan metabolism will be dealt with in the next chapter.

3. Lactose metabolism in *E. coli*

The first detailed study of metabolic regulation was the degradation of the sugar lactose to galactose and glucose in the bacterium *E. coli*. These studies were performed as early as the late 1950s. Lactose and other β-galactosides are metabolised in *E. coli* by the enzyme β-galactosidase. The enzyme is very specific and acts only on β-galactosides that are unsubstituted on the galactose ring.

In *E. coli* there is only one type of β-galactosidase and the organism does not have any other enzyme capable of metabolising lactose.

In *E. coli*, lactose is metabolised by the enzyme β-galactosidase. The bacterium only expresses the β-galactosidase gene when lactose is available as a carbon source.

In the presence of lactose, β-galactosidase is strongly induced.

However, the possession of β-galactosidase activity alone is not sufficient to allow *E. coli* to metabolise lactose. At least one other component, separate and distinct from β-galactosidase, is required and this is used to transport lactose into the bacterial cell.

E. coli can use lactose as its sole carbon source. An *E. coli* bacterium growing on lactose will contain several thousand molecules of the β-galactosidase enzyme. In contrast, a bacterium growing on other carbon sources, such as glycerol or glucose, will have fewer than ten copies of this enzyme.

The presence of lactose in the culture medium induces a large increase in the production of active β-galactosidase; that is, β-galactosidase is an inducible enzyme. When growing in the presence of lactose, two other proteins are also produced by the bacterium. These proteins are known as lactoside permease and thiogalactoside transacetylase. The permease is required for the transport of lactose across the bacterial cell membrane and is necessary for lactose metabolism. The *in vivo* role of the transacetylase enzyme remains unclear; *in vitro*, it can catalyse the transfer of an acetyl group from acetyl-coenzyme A to the C-6 hydroxyl group of thiogalactoside.

That the level of β-galactosidase changes dramatically in the presence of lactose illustrates that the enzyme is under strict control. Another feature of this system is its extreme specificity; only substrates of β-galactosidase or substances very closely related to substrates act as inducers. A product of lactose, allolactose, is the actual physiological inducer of protein synthesis from the *lac* operon. Allolactose is formed from lactose by transglycosylation. Studies of synthetic inducers have shown that some β-galactosides are inducers without being substrates of β-galactosidase, whereas other compounds are substrates without being inducers. For example, isopropylthio-galactoside (IPTG) is a non-metabolisable inducer and X-gal is a substrate but is unable to induce β-galactosidase synthesis. The chemical structures of lactose, allolactose and IPTG are shown in Figure 9.1.

4. THE LACTOSE OPERON

Nobel Laureates Francois Jacob and Jacques Monod conducted most of the work on the lactose operon structure and its control mechanisms. Using specific, mutated *E. coli* lines, they investigated the mechanisms involved in β-galactosidase control.

An initial observation in the development of the operon hypothesis for lactose degradation came from the finding that the levels of the permease and transacetylase increase in direct proportion to the level of β-galactosidase. Specifically, the ratio of the three enzymes was 10 (β-galactosidase): 5 (permease): 2 (transacetylase), under all circumstances, immediately suggesting a common control mechanism.

4.1 Jacob & Monod's mutants

By isolating and studying *E. coli* strains with specific mutations, Jacob and Monod showed that the β-galactosidase, the permease, and the

a)

b)

c)

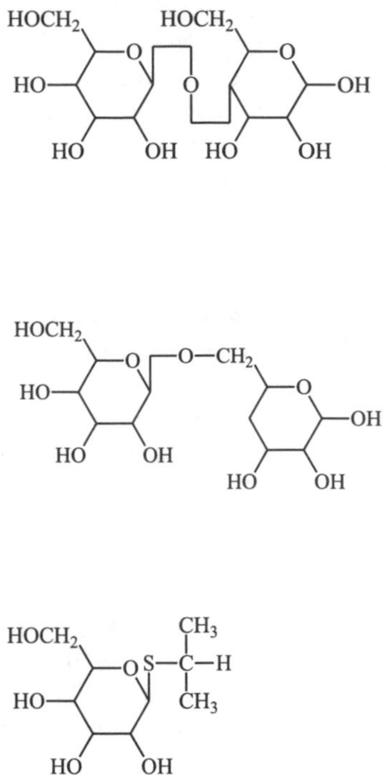

Figure 9.1 Structures of (a) lactose, (b) allolactose, and (c) IPTG.

transacetylase were encoded by three contiguous genes, named *lac z*, *lac y*, and *lac a*, respectively.

Nobel Laureates Francois Jacob and Jacques Monod studied the lactose operon using dozens of mutant bacteria that carried deficiencies or abnormalities in lactose metabolism.

Mutants containing the active form of two of these three proteins were isolated. For example, the nomenclature *z–y+a+* denotes a mutant lacking *β*-galactosidase (*z–*) but containing active permease (*y+*) and transacetylase (*a+*). A second class of mutant isolated contains all three proteins in their active form, that is *z+y+a+* bacteria. These mutants produce the three proteins in large amounts regardless of whether any inducer is present. These mutants are known as constitutive mutants.

Jacob and Monod deduced that "the rate of synthesis of these three proteins is governed by a common element that is different from the genes specifying their structure". They named the gene for this common regulatory element *I*. Wild-type, normally inducible bacteria have the genotype *i+z+y+a+*. The protein product of the *I* gene acts as a repressor and controls the expression of the *z*, *y*, and *a* genes.

Constitutive lactose mutants can be divided into two subgroups, O^c or I^-. O^c mutants constitutively express the enzymes and contain changes in the operator (binding site for *I* gene product), but still produce an active product from the *I* gene (repressor). I^- mutants do not produce an active form of the repressor protein, but do contain the correct operator sequence (repressor binding site).

The question of how the product from the *I* gene acts to control the production of *β*-galactosidase, permease, and transacetylase was answered by a set of simple but ingenious experiments using the mutants already described. The work also illustrates hypothesis testing, experimental design, and logical, deductive analysis of results.

The simplest hypothesis was that the product from the *I* gene acted as a repressor controlling the production of the three proteins. The experiments conducted used cells containing two copies of each gene, with at least one being active, and relied not on molecular biology techniques, but on bacterial physiology to accomplish this. The first copy in each case was located on the bacterial chromosome; the second copy was introduced into the bacterium on a plasmid (an extrachromosomal piece of DNA) by a process known as F-duction.

Once the plasmid is introduced into the cell, the cell is effectively diploid for the genes in question. By combining different mutant strains and different F′-plasmids, Jacob and Monod were able to form different combinations of native and mutated genes and to deduce how the system operated.

In the rest of this chapter we will first describe how the lactose operon is now known to be oriented and controlled and then explain the experimental design that led to the elucidation of this mechanism.

4.2 The structure and mechanism of the lactose operon

An "operon" is a set of functionally related bacterial genes under common control. The lactose operon of *E. coli* encodes proteins essential to the bacterium in metabolising lactose.

In contrast to the control mechanisms seen in eukaryotes, *negative* control of expression of bacterial operons is the norm.

In bacteria, genes encoding proteins involved in the same process are often found immediately adjacent to each other; in addition to being physically close or adjacent, regulation of expression of these genes is such that they are all turned on or off together, that is, they are regulated in a co-ordinated manner. Such a group of co-ordinately regulated genes, together with the control elements, is referred to as an operon. Such a grouping of related genes under a common control mechanism permits the bacteria to adapt rapidly to changes in the environment, including the availability of a carbon source or the absence of a necessary amino acid.

The lactose operon of *E. coli* encodes proteins involved in the metabolism of lactose. Under normal conditions the operon is only induced when lactose is available. The operon itself is known to contain three active genes denoted by the letters *Z*, *Y* and *A*. The *Z* gene encodes β-galactosidase, the *Y* gene encodes a permease, and the *A* gene encodes a transacetylase enzyme. These genes lie adjacent to each other in the *E. coli* genome in the order 5′–*Z*–*Y*–*A*. Preceding these genes there are two other important regions, the operator (O) and the promoter (P) (indicated in Figure 9.2). The operator acts as a switch for the activation of the operon and the promoter is the site from which gene transcription starts.

In contrast to the control mechanisms seen in eukaryotes (discussed in chapter 4), *negative* control of expression of bacterial operons is the norm. Frequently, a negatively acting protein called the repressor regulates expression of an operon by binding to a DNA site, referred to as an operator, near the promoter for transcription. Expression of repressor protein is usually not controlled by a repressor; the repressor gene is constitutively active at some low level. The operator is a

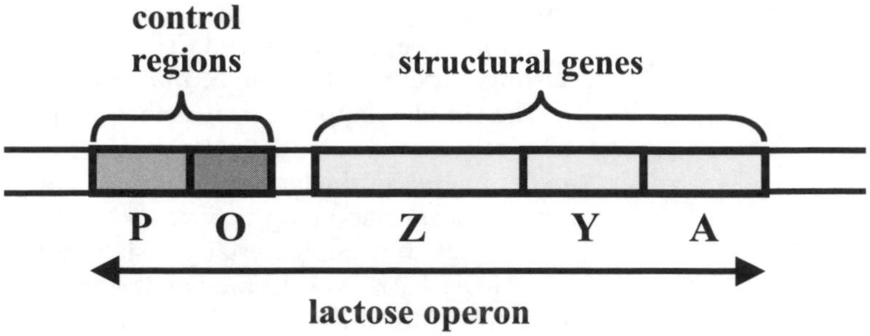

Figure 9.2 The lactose operon and its constituent structural genes. Z = β-galactosidase gene, Y = permease gene, A = acetylase gene, O = operator sequence, P = promoter sequence.

binding site for the repressor, a negative transcription factor. Frequently, repressor activity is regulated or influenced by a small molecule, usually a metabolite in the pathway in question.

A gene not present in the lactose operon, known as *I*, encodes a protein that binds to the operator sequence and represses the operon, that is, suppresses transcription. Thus, in the "normal" state, in which lactose is not the bacterium's carbon source, the operon is inactive because the repressor binds to the operator and occludes it.

In the absence of lactose in the medium, the operon is negatively controlled and the level of β-galactosidase within the bacterium is extremely low (fewer than 10 molecules of enzyme in the bacterium). In the presence of lactose, however, the level of active β-galactosidase enzyme increases to over a thousand molecules. The presence of lactose eliminates the suppression caused by the *lac* repressor. Lactose actually binds to the repressor, causing some conformational change after which the binding affinity of the repressor for its binding site is reduced and the repressor then dissociates. The initiation of transcription can then proceed (Figure 9.3).

As is the norm in bacteria, the messenger RNA (mRNA) that is transcribed encodes all three genes in one continuous message. The structure and mechanism of action of the lactose operon was determined using a series of mutant bacteria with mutant lactose operons.

4.3 Experimental design

One of the first questions to be answered concerning the *lac* operon was whether the increase in β-galactosidase, permease, and transacetylase proteins resulted from new protein synthesis or the activation of pre-existing protein precursors. Using immunological and isotopic labelling techniques it was readily shown that the increased level of β-galactosidase was the result of new protein synthesis.

Another preliminary issue was whether the permease was truly a different molecule to the β-galactosidase enzyme, for at least theoretically it was possible that the β-galactosidase enzyme acted as a permease and allowed lactose to cross the bacterial cell wall. This

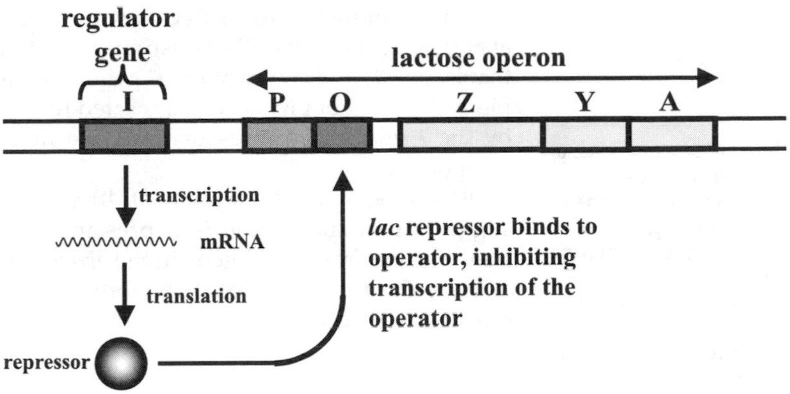

Figure 9.3 Mechanism of action of the *lac* repressor.

question was settled by isolating *E. coli* mutants that had lost the ability to form active β-galactosidase enzyme but retained the ability to concentrate lactose in the cytoplasm. Additionally, the converse mutants were isolated; these bacteria formed active β-galactosidase but lactose entered the cell to a very low degree (these mutants were referred to as cryptic mutants).

Information regarding the physical positioning of the three structural genes of the operon came from genetic mapping of the many mutations present in mutant *E. coli*. All the mutations mapped to a very small region of the bacterial chromosome, now known as the *lac* region. In addition, this mapping exercise gave the relative positions of the three genes and the likely positions of the operator (*O*) and the repressor gene, *I*.

4.4 The lactose repressor

Mutations within the *I* gene give rise to an altered *lac* repressor protein; such mutations can cause a number of effects. First would be a mutation that changed an amino acid residue in the protein without affecting its overall structure or its function. Such "silent" mutations would go unnoticed (in the days before DNA sequencing!). Second would be mutations that stop the *lac* repressor from inhibiting transcription from the operon (so-called I^- mutants). Such I^- mutants either do not express the lac repressor at all or the protein that is expressed is inactive (*i.e.*, does not bind DNA). I^- mutants are constitutive mutants; that is, they express *lac* operon proteins *regardless* of the presence or absence of lactose in the medium. In fact, most I^- mutants actually synthesise *more* β-galactosidase at all times than do induced wild-type cells. However, while the level is altered, the *ratio* of the β-galactosidase to permease to transacetylase is the same in the constitutive cells as in induced wild-type cells, again pointing strongly to a common control mechanism.

Next, examination of merodiploid mutants carrying the genotypes I^+Z^-/F I^-Z^+ (that is I^+Z^- genes on the F plasmid, I^-Z^+ genes in the bacterial chromosome) and I^-Z^+/F I^+Z^- showed that these bacteria were both inducible for active enzyme, not constitutive. Thus, the I^+ allele is dominant over the constitutive mutant and it is active in *trans;* that is, one "good" copy of the I^+ gene can effect regulation of another piece of DNA not physically connected to it. Thus, the control effected by the *I* gene (*lac* repressor) must be mediated by a cytoplasmic component.

The *I* gene encodes a protein that can repress transcription from the Z gene in the *absence* of lactose. In the *presence* of lactose, this protein loses its ability to inhibit transcription.

Of course, with the benefit of 40 years of hindsight this result suggests that the *I* gene is expressing a protein that can repress transcription from the Z gene in the absence of lactose. On addition of lactose, the repressor loses its ability to inhibit transcription and transcription occurs.

4.4.1 *The "super" repressor*

This hypothesis was confirmed using another kind of *I* mutant, referred to as I^S, the so-called "super" repressor. Such mutants have completely lost their ability to synthesis *β*-galactosidase, permease, or transacetylase. Furthermore, it can easily be shown that this is not simply the result of deletion of these genes or their promoter. Specifically, if I^S bacteria are recombined with Z^-Y^- mutants, the resultant bacteria *can* be induced by lactose, demonstrating that the *Z* and *Y* genes are present but are being repressed in the I^S mutant. Using merodiploids (I^S/I^+), it was clear that I^S was dominant; *i.e.*, those merodiploids cannot synthesise *β*-galactosidase or acetylase.

Again with hindsight, the most consistent way of explaining this type of mutant would be to posit that the *lac* repressor expressed in such bacteria can still associate with its binding site as tightly as the wild-type repressor, but that it has lost its ability to be inactivated by lactose. Presumably, in the I^S mutant, the mutations are such as to alter the binding site for lactose so that it no longer binds or prevents the conformational change when it does bind.

In "super repressor" I^S mutants, the repressor continues to block transcription of the Z gene even *in the presence* of lactose.

If this explanation is correct, then further mutations within the *lac* repressor may turn the I^S cells into constitutive mutants, regaining their ability to metabolise lactose. In these experiments, 50% of the "revertants" become I^-. The other 50% become constitutive mutants, termed O^C (see below).

From this evidence, Jacob and Monod deduced that a protein was being produced from a gene and this protein's function was to control the other genes. Such a regulator gene does not contribute structural information to the proteins that it controls. The product of the regulator gene is a cytoplasmic substance (today with hindsight, obviously a protein, and more than that, a transcription factor) that controls the transcription of mRNA from genes.

4.5 The operator

From the discussion so far, it should be apparent that the *Z*, *A*, and *Y* genes in *E. coli* are controlled by the lac repressor, itself encoded by the *I* gene. How then does the *lac* repressor control these genes?

Once a cytoplasmic repressor that can act on more than one gene is envisaged, the existence of an operator element (*i.e.*, a binding site for the repressor) must follow. Additionally, once a binding site is envisaged, it is clear that mutations within such a site might stop the binding of the repressor.

The key property of the repressor was its ability to act on more than one gene. Additionally, the repressor is both highly specific, in that mutations in the *I* gene do not affect any other system, and pleiotropic, in the sense that β-galactosidase, permease, and transacetylase are affected simultaneously and to the same extent, by such mutations.

Once it was shown that the repressor existed and was cytoplasmic, the next question was its site of action. In the *lac* operon this controlling element is known as the "operator". In fact, once the existence of a specific repressor is considered established, the existence of an operator element (a binding site for the repressor) must follow.

Once an operator binding site (on the DNA) is envisaged, it is clear that mutations within such a site might stop the binding of the *lac* repressor. Such mutations, however, would not necessarily affect the bacterium's ability to initiate transcription and thus produce β-galactosidase. In fact, such mutations would result in a constitutive phenotype and would be predicted to be dominant over the wild-type organism.

In the *lac* system, dominant constitutive (O^C) mutants were isolated by selecting for constitutivity in bacteria diploid for the *lac* region. Using merodiploid bacteria, I^- mutants are virtually eliminated, because for the bacteria to be functionally I^-, both copies of the *I* gene need to be mutated within the same bacterium. However, for the O^C genotype only one form of the operator needs to be genetically altered. By recombination, the O^C mutations were mapped in the *lac* region, between the *I* and the *Z* loci, the order being *I–O–Z–Y–A*.

If the constitutivity of the O^C mutant results because the *lac* repressor can no longer bind to the operator because of mutations in the operator, the O^C organisms should be insensitive to the presence of the altered repressor synthesised by the I^S gene. Thus, in merodiploid bacteria with the genotype I^SO^+/F′ I^+O^C, the prediction is that the bacteria will still be constitutive. This was found to be the case and the observation confirms the interpretation of the effects of both the mutations, I^S and O^C. In addition, consistent with this analysis, O^C mutants frequently arise as lactose positive "revertants" in populations of I^S bacteria.

As with I^- mutants, O^C mutants are pleiotropic; the mutation(s) affects – simultaneously and quantitatively to the same extent – the synthesis of β-galactosidase, permease, and transacetylase. This result again points to a single operator controlling all three genes.

This proposed single operator controlling the operon genes was indeed identified using a variety of merodiploid mutants. Using the mutants O^+Z^+/F′ O^CZ^- and O^+Z^-/F′ O^CZ^+ both the normal β-galactosidase and a mutated version of β-galactosidase are produced in the presence of lactose. However, in the absence of inducer, only the protein corresponding to the Z allele in a *cis* position to O^C is produced. O^C therefore has no effect on the Z allele in the *trans* orientation. This indicates that the operator O^C does not produce a cytoplasmic product but is a DNA sequence, the binding site of the *lac* repressor.

Additionally, certain mutations within the operator region can modify the operator in such a way as to inactivate the *lac* operon, resulting in the loss of the capacity to synthesise β-galactosidase. Such O^O mutants are recessive to O^+ or O^C, again demonstrating that the operator is a binding site for the *lac* repressor and does not itself produce a diffusible product. In these mutants, the *I* gene is functional, showing clearly that not only are the *I* and *O* mutants not alleles, but that the *O* segment, while governing the expression of the *Z*, *Y*, and *A* genes, does not affect the expression of the regulator gene, which is located outside the operon.

From the evidence presented so far, it seems that the lactose operon is a collection of three consecutive genes all controlled by a repressor molecule binding to a single site, the operator. Under normal environmental conditions the operon is repressed. However, under certain conditions (the presence of lactose), this repression is lifted, allowing production of β-galactosidase, the enzyme which metabolises lactose (Figure 9.4).

5. CONFIRMATION OF THE LACTOSE OPERON MODEL

The lactose operon model as first proposed by Jacob and Monod was later studied and confirmed by Gilbert and Müller-Hill, with a series of further experiments leading to the isolation and characterisation of the *lac* repressor.

5.1 Isolation of the *lac* repressor

The initial isolation of the *lac* repressor was carried out using a technique known as equilibrium dialysis. By using a radiolabelled mimic

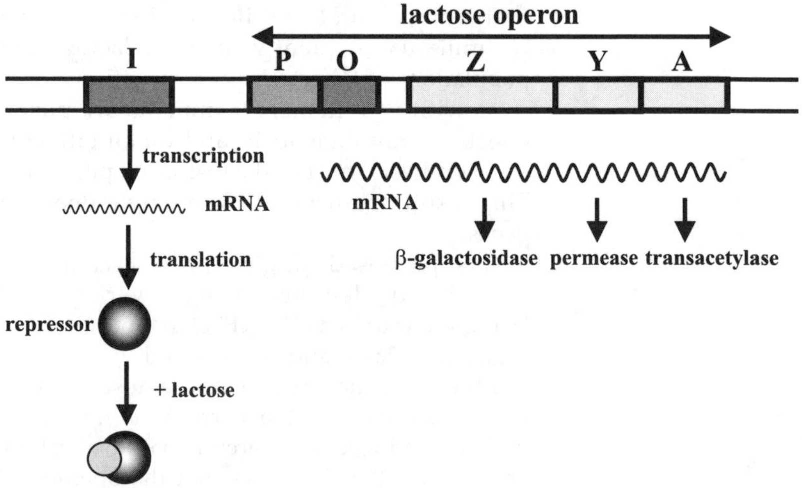

Figure 9.4 The essential elements of the *lac* operon and its regulatory mechanism.

of lactose, ^{14}C-IPTG, Gilbert and Müller-Hill were able to isolate the *lac* repressor. IPTG binds to the *lac* repressor as an inducer but it is not a substrate of β-galactosidase, and so is not metabolised.

The dialysis bag contains fractions of *E. coli*. The buffer contains ^{14}C-IPTG. The IPTG can pass through the dialysis bag's membrane whereas proteins and other large molecules cannot.

If a particular bacterial fraction placed in the dialysis bag contains a substance that binds the radioactively labelled lactose mimic (^{14}C-IPTG), then ^{14}C-IPTG will become concentrated inside the dialysis bag. Such an increased concentration is readily detected, because the IPTG is radioactively labelled (^{14}C-IPTG) (Figure 9.5).

From this equilibrium dialysis experiment, Gilbert and Müller-Hill were able to deduce which cellular fraction of *E. coli* contained the *lac* repressor. Of course it was possible that *E. coli* contains more than one protein that binds IPTG (and increases the concentration of IPTG in the bag). To assess this problem Gilbert and Müller-Hill compared their results with the *E. coli* mutant I^S. They found that the fraction from I^S that corresponded to the fraction from wild-type *E. coli*, which concentrated IPTG, had no effect on the IPTG concentration. This confirmed the notion that the I^S super-repressor was "immune" to the effect of lactose (or IPTG) because lactose (or IPTG) failed to bind to it. The I^S mutant therefore acted as a control and confirmed that Gilbert and Müller-Hill had isolated the *lac* repressor protein.

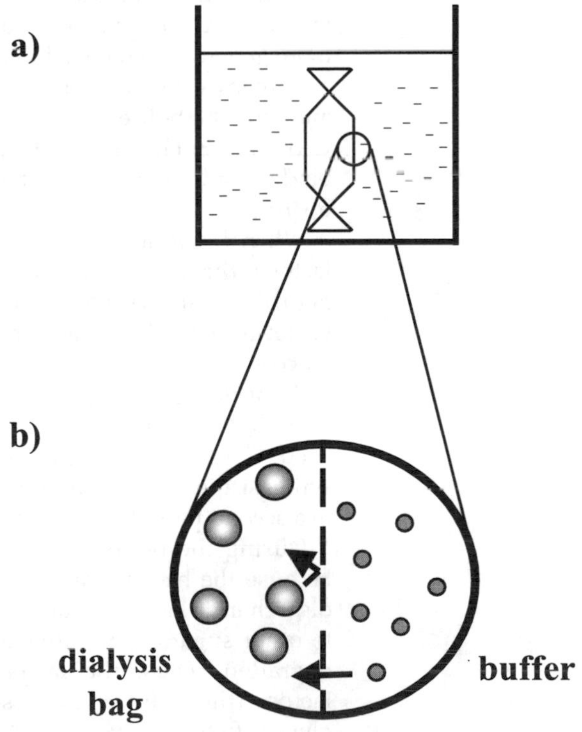

Figure 9.5 Equilibrium dialysis.

5.2 Characterisation of the operator

Smith and Snyder characterised a further group of O^C mutants in more detail. All mutants made β-galactosidase in the absence of lactose but at *differing* levels. Following the development of DNA sequencing, it was discovered that all these mutations were located within the *O* region, a very small region of DNA immediately 5′ to the *Z* gene, and most were single base substitutions.

A possible explanation for why these single base substitutions led to different levels of β-galactosidase in the absence of lactose is that they affect the efficiency with which RNA polymerase binds to the operator sequence. The key here is that the various O^C mutations lead to *differing* levels of β-galactosidase; it was not all-or-nothing, as was the I^S/I^+ paradigm.

6. CATABOLITE REPRESSION, OR THE "GLUCOSE EFFECT"

When *E. coli* are grown in medium containing both lactose and glucose, the lactose operon is not activated until all the glucose has first been used. This phenomenon is referred to as the "glucose effect" because glucose repressed the synthesis of certain inducible enzymes, including those of the lactose operon, even if lactose was present in the medium. A similar process occurs in other operons and the process more generally is referred to as catabolite repression.

In many catabolic operons, a complex of 3′, 5′-cyclic AMP (cAMP) and the catabolite activator protein (CAP) binds to the so-called CAP site. Such a site is located in the *lac* operon promoter. This binding of CAP + cAMP is necessary for expression of these catabolic operons.

When *E. coli* are grown in medium containing both glucose and lactose, the bacteria exhibit a biphasic growth curve. Graph a) of Figure 9.6 shows bacterial growth on glucose. Graph b) shows bacterial growth on lactose. Graph c) shows bacterial growth on glucose + lactose.

What is happening on graph c? In the first period of exponential growth, the bacteria preferentially metabolise glucose, until it is all used. Then, after a lag period, during which time the bacteria are synthesising enzymes to metabolise lactose, the lactose is metabolised in a second period of exponential growth.

During the period of glucose utilisation, lactose is not utilised because the bacteria are unable (as yet) to transport and cleave the disaccharide lactose. Glucose is always metabolised first in preference to other sugars. Only after glucose is completely utilised is lactose degraded. Thus, the lactose operon is repressed even though lactose (the inducer) is present. It is simply more efficient to use glucose first.

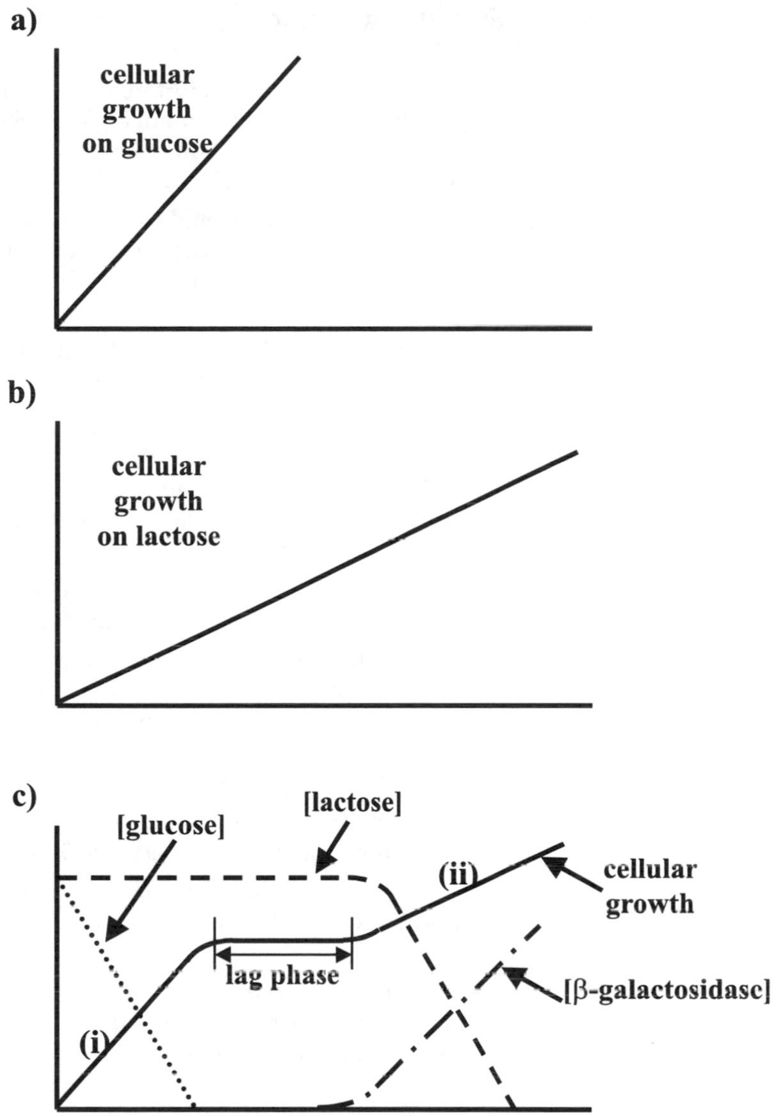

Figure 9.6 Biphasic growth curve. The bacteria will grow exponentially on glucose (graph A) or lactose (graph B). When glucose *and* lactose are *both* present in the bacterial environment, the bacteria will preferentially use glucose. As indicated on graph C, the glucose concentration decreases as the cells grow (a), while the lactose concentration remains constant. When all the glucose has been consumed, β-galactosidase synthesis will be induced. During this period, cellular growth will slow, or stop, the "lag phase". As the β-galactosidase concentration increases, the cells will be able to utilise lactose. Consequently, the levels of lactose will begin to fall and cellular growth will resume (b), although at a slower rate than when metabolising glucose.

Summary

The lactose operon in *E. coli* contains three genes that are required for lactose metabolism. The genes are expressed co-ordinately. The operon is ordinarily switched off and the bacterium will preferentially use glucose even if lactose is present in the medium; this is mediated at the level of transcription. When, however, lactose is the only carbon source available, the bacterium induces transcription of the enzymes of the operon. Lactose actually binds to and causes a conformational change in the repressor protein that ordinarily switches off the operon.

Further reading

1. Lewis, M., Chang, G., Horton, N. C., Kercher, M. A., Pace, H. C., Schumacher, M. A., Brennan, R. G. and Lu, P. (1996) Crystal structure of the lactose operon repressor and its complexes with DNA and inducer. *Science* **271**, 1247–1254.
2. Hediger, M. A., Johnson, D. F., Nierlich, D. P. and Zabin, I. (1985) DNA sequence of the lactose operon: the lacA gene and the transcriptional termination region. *Proc. Natl. Acad. Sci. USA* **82**, 6414–6418.
3. Malan, T. P. and McClure, W. R. (1984) Dual promoter control of the *Escherichia coli* lactose operon. *Cell* **39**, 173–180.
4. Betz, J. L. and Sadler, J. R. (1976) Tight binding repressors of the lactose operon. *J. Mol. Biol.* **105**, 293–319.
5. Gilbert, W., Maizels, N. and Maxam, A. (1973) Sequences of controlling regions of the lactose operon. *Cold Spring Harb. Symp. Quant. Biol.* **38**, 845–855.
6. Reznikoff, W. S. (1992) The lactose operon-controlling elements: a complex paradigm. *Mol. Microbiol.* **6**, 2419–22.
7. Amouyal, M. and von Wilcken-Bergmann, B. (1992) Repression of the *E. coli* lactose operon by co-operation between two individually unproductive "half-operator" sites. *C. R. Acad. Sci.* **315**, 403–407.
8. Yu, X. M. and Reznikoff, W. S. (1986) Deletion analysis of RNA polymerase interaction sites in the *Escherichia coli* lactose operon regulatory region. *J. Mol. Biol.* **188**, 545–553.
9. Leive, L. and Kollin, V. (1967) Synthesis, utilisation and degradation of lactose operon mRNA in *Escherichia coli*. *J. Mol. Biol.* **24**, 247–259.
10. Epstein, W., Naono, S. and Gros, F. (1966) Synthesis of enzymes of the lactose operon during diauxic growth of *Escherichia coli*. *Biochem. Biophys. Res. Commun.* **24**, 588–592.

End of unit questions

1. How did the super repressor, I^S, mutants further our understanding of the operation of the lactose operon?
2. What is the role of β-galactosidase in lactose metabolism?
3. Explain equilibrium dialysis and its use in the study of the lactose operon.

PROBLEMS

1. Explain the expression of lactose operon enzymes in a bacterium growing in a medium containing both glucose and lactose (see reference 10).
2. Discuss how the structure of the lactose repressor protein is related to its function (see reference 1).

10 Other Transcription Systems, Eukaryotic and Prokaryotic

OBJECTIVES

a) to introduce a series of other well-characterised transcription systems in eukaryotes and prokaryotes

1. Overview

This chapter describes a series of further transcription systems, both eukaryotic and prokaryotic, to be compared and contrasted with the systems previously described. As will be seen, the systems share many fundamental features.

2. The tryptophan operon in *E. coli*

The tryptophan operon in *E. coli* is widely considered to be the paradigmatic anabolic (biosynthetic) operon in bacteria; it has been heavily studied and is well understood. The operon encodes the enzymes necessary for tryptophan biosynthesis. The pathway is regulated at several levels, including at the level of transcription. The mechanism outlined here shares many features with the lactose operon described in the previous section, but also includes several distinct and unique elements.

Genes encoding the enzymes for tryptophan biosynthesis in *E. coli* are organised in an operon, under common control.

The genes encoding the enzymes for tryptophan biosynthesis in *E. coli* are organised in an operon, under common control. Specifically, the operon comprises a promoter (*P*) region, an operator (*O*) sequence, an attenuator (*A*) region, and five structural genes encoding the enzymes (referred to as *Trp A–E*) (Figure 10.1a). The attenuator sequence is located between the operator and the structural genes and constitutes a further "barrier" that RNA polymerase must traverse before transcribing the structural genes. When tryptophan is present in the medium, most RNA polymerase molecules do not get as far as transcribing the genes. When none is present, RNA polymerase successfully transcribes the genes.

Additionally, there is a "regulator" gene, separate from the operon, which encodes the tryptophan repressor protein. This repressor is inactive unless tryptophan is bound to it. Active repressor binds to the operator region and blocks binding of RNA polymerase, preventing transcription. As a result, none of the enzymes required for the biosynthesis of tryptophan is synthesised (Figure 10.1b). The repressor is synthesised constitutively at a low level.

2.1 Attenuation

Attenuation is an additional regulatory mechanism, present in the tryptophan and several other operons. There is a built-in transcription "terminator" near the beginning of the operon. In detail, RNA polymerase assembles on the promoter and initiates transcription. A short region of DNA is transcribed, and the resulting mRNA encodes a leader peptide of 14 amino acids, two of which are tryptophan residues.

The transcribed attenuator mRNA sequence can form two alternative hairpin structures. When regions 3 and 4 pair – by hydrogen bonding – transcription termination results (Figure 10.2a).

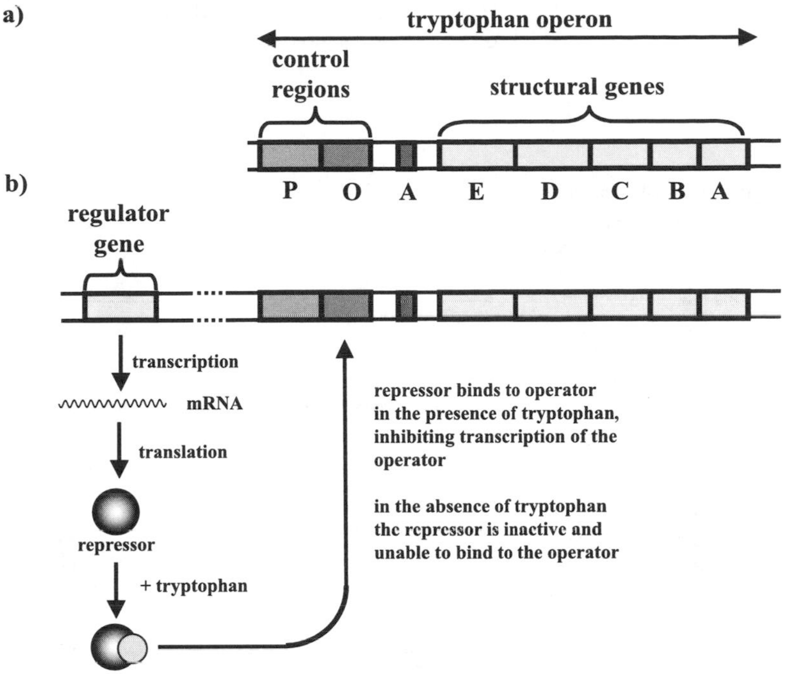

Figure 10.1 The tryptophan operon of *E. coli.*
(a) Organization of the tryptophan operon; the operon comprises a promoter (*P*) region, an operator (*O*) sequence, an attenuator (*A*) region, and five structural genes encoding the enzymes (referred to as *Trp A-E*)
(b) Method of operation; when tryptophan is present in the medium, most RNA polymerase molecules do not get as far as transcribing the genes. when none is present, RNA polymerase successfully transcribes the genes.

2.2 How does this hydrogen-bonded hairpin lead to a regulatory mechanism?

Suppose there is no tryptophan in the medium and bacterium needs to biosynthesize the amino acid. Then, in absence of tryptophan, the ribosome attempting to translate the newly transcribed mRNA will stall at the two tryptophan codons, there being no tryptophan-tRNA available, allowing regions 2 and 3 to pair. If region 3 is bound to region 2, it cannot bind to region 4, and without the 3 to 4 hairpin there is no transcription termination, and the structural genes of the operon are transcribed (Figure 10.2b). Thus, the enzymes necessary for tryptophan biosynthesis will be synthesised in the absence of tryptophan.

Regulation of the tryptophan operon depends on the uniquely prokaryotic phenomenon of *coupled* transcription and translation.

If tryptophan is present, then the ribosome will have no problem translating the leader peptide. In moving through the two tryptophan codons, the 2 to 3 hairpin will be disrupted, allowing 3 to 4 pairing, causing transcription termination. Thus, the enzymes necessary for tryptophan biosynthesis will not be synthesised in the presence of tryptophan.

Thus, in a mechanism uniquely prokaryotic, *coupled* transcription and translation is crucial. Termination results from a stem-loop structure formed in the mRNA, *not* in the DNA. RNA polymerase pauses after the structure.

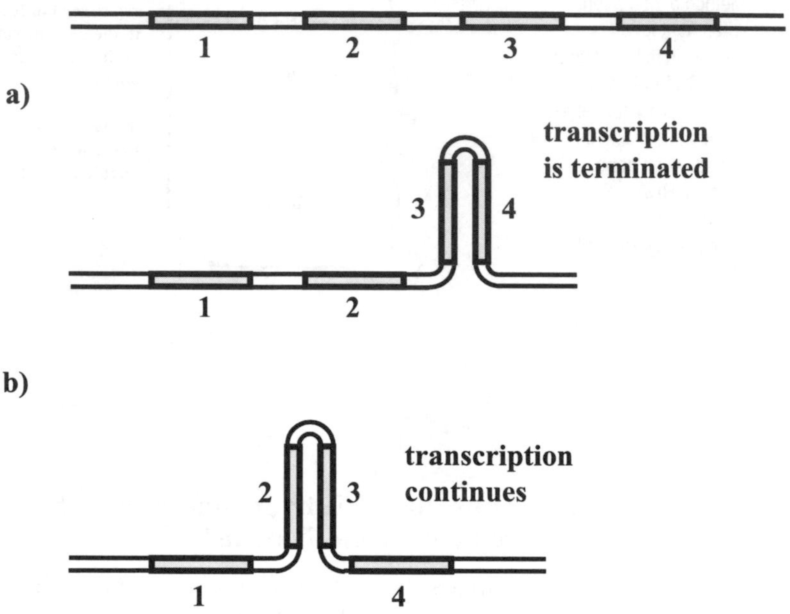

Figure 10.2 Attenuation.
(a) If tryptophan is present, the ribosome will have no problem translating the leader peptide. In moving through the two tryptophan codons, the hairpin between regions 2 and 3 will be disrupted, allowing regions 3 and 4 to pairing, causing transcription termination. Thus, the enzymes necessary for tryptophan biosynthesis will not be synthesised in the presence of tryptophan.
(b) When there is no tryptophan in the medium, the bacterium needs to biosynthesise it. When the ribosome attempts to translate the newly transcribed mRNA, it will stall at the two tryptophan codons, there being no tryptophan-tRNA, allowing regions 2 and 3 to pair. If region 3 is bind to region 2, it cannot bind to region 4 and without the 3 to 4 hairpin, there is no transcription termination, and the structural genes of the operon are transcribed.

3. The human β-globin gene

The human β-globin gene has been widely studied for many years as a model transcriptional system, and involves a complex web of several interacting mechanisms. The pattern of globin gene expression varies in a tissue-specific and developmental stage-specific manner. As an example, there is a switch from human foetal γ-globin to adult β-globin expression, within erythroid precursor cells.

The key DNA sequence, in terms of globin expression, is the so-called "locus control region" (LCR), which is located far upstream of the β-globin gene itself. Locus control regions (LCRs) are *cis*-acting DNA segments needed for activation of an entire locus or gene cluster. The LCR is believed to alter chromatin structure, and to affect transcription of the β-globin genes. Many transcription factors interact with the LCR and the individual promoters and enhancers of the β-globin genes. The LCR alters chromatin structure, rendering the entire β-globin locus susceptible to nucleases. It is thought that this nuclease-sensitive state permits transcription of the globin genes. However, the exact details of this regulation remain unknown.

The human β-globin gene is largely regulated by the so-called "locus control region" (LCR), located distant from the gene.

The human β-globin LCR contains five DNase I hypersensitive (HS) sites located 5′ to the multiple genes it regulates, 5′HS1–5. Deletion of individual sites decreases the ability of the LCR to bring about the open chromatin conformation. Presumably, then, these five sites interact co-operatively.

4. Phosphate metabolism in yeast

In order to integrate growth and division with the nutritional status of the cell there must be communication between many different pathways, resulting in a co-ordinated control of transcription in response to changing environmental factors. A well-studied example of such transcriptional control is that of phosphate metabolism in yeast.

The yeast *Saccharomyces cerevisiae* has at least six types of phosphatases, found at different locations throughout the cell, and inorganic phosphate transporters. The genes encoding these enzymes are repressed and de-repressed in a co-ordinated manner, depending on the availability of phosphate in the growth medium.

The induction of phosphatase enzymes in yeast in conditions of phosphate starvation is mediated at the level of transcription by helix-turn-helix and homeodomain proteins.

If inorganic phosphate becomes scarce, yeast will react by co-ordinating the induction of the phosphatase genes. The product of one of these genes, known as PHO5, can act as an extracellular phosphate scavenger. Under high phosphate conditions there is no need for the cell to produce large amounts of PHO5 and the gene is repressed by a variety of proteins. Several genes have been identified as being essential for the correct regulation of the PHO5 gene.

Expression of PHO5 is strongly induced in conditions of phosphate starvation. The actions of the helix-turn-helix transcription factor Pho4p and the homeodomain-containing protein Pho2p mediate this induction. For the protein Pho4p, there are two binding sites within the promoter, P1 (low affinity binding) and P2 (high affinity binding).

Under low phosphate conditions, repression of *Pho5* is lifted and the gene is expressed. Activation of the *Pho5* gene (low phosphate conditions) requires an additional protein, Pho81p. The structure of Pho81p suggests that it can form strong protein–protein interactions. Under high phosphate conditions (repression of *Pho5*), Pho81p interacts with Pho80p, and Pho80p associates with Pho85p and Pho4p. In low phosphate (activation of *Pho5*) conditions, Pho80 and Pho81 continue to interact with each other but dissociate from Pho4p. Following this dissociation, Pho4p can associate with Pho2p and bind co-operatively at the *Pho5* promoter, activating transcription initiation.

5. THE STRINGENT RESPONSE

The so-called "stringent response" is a co-ordinated process in bacteria, typically triggered by amino acid starvation, which leads to a huge change in metabolism. As 50% (w/w) of total cell mass is protein, it should be clear that protein biosynthesis requires a considerable fraction of all cellular resources.

When bacteria growing in glucose minimal medium containing all amino acids are then transferred to the same medium containing no amino acids, the stringent response is induced. In this response, the concentration of charged tRNAs falls, because there are suddenly no amino acids to charge the tRNAs, and two highly phosphorylated nucleotides, guanosine tetraphosphate (5′-ppGpp-3′) and guanosine pentaphosphate (5′-pppGpp-3′), are synthesised (Figure 10.3).

Protein synthesis decreases and there is a switch to producing mainly the enzymes of the amino acid biosynthetic pathways, such as those of the tryptophan operon discussed earlier in this chapter. The synthesis of all other non-essential biomolecules is greatly reduced, including that of rRNA, tRNA, phospholipids, and carbohydrates. As the enzymes of amino acid biosynthesis begin to make amino acids, the level of charged tRNAs rises again and the bacterium adapts to a new slower rate of metabolism and growth in the new nutrient-poor medium.

Mutant bacteria unable to mount a stringent response or unable to do so normally are known, and are referred to as "relaxed" mutants. Such bacteria do not accumulate (p)ppGpp.

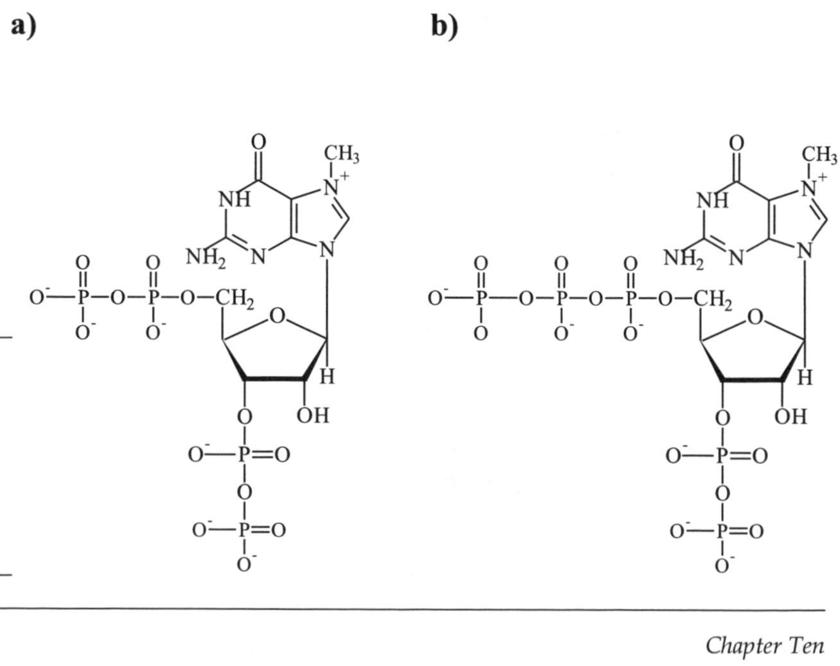

Figure 10.3 Chemical structures of (a) guanosine tetraphosphate (5′-ppGpp-3′) and (b) guanosine pentaphosphate (5′-pppGpp-3′).

RelA is a protein associated with the ribosome. When the concentration of an amino acid falls, so does the amount of charged tRNA. As a result, ribosomes stall when they are unable to extend a protein beyond the "missing" amino acid. Such stalling activates RelA, a protein with enzymatic activity. Specifically, RelA synthesises ppGpp from GDP and ATP or pppGpp from GTP and ATP. There also appears to be a relA-independent pathway in that some relA$^-$ mutants can also mount a stringent response.

The exact role of these transient molecules is currently unclear, but it is believed that ppGpp can interact with the β subunit of RNA polymerase and apparently alters the polymerase's promoter specificity. Transcription of some operons is inhibited, while at others it is stimulated.

6. Heat shock

When cell or tissue injury occurs, the inflammatory response is triggered, largely through the action(s) of arachidonic acid and its metabolites. As a result, there is expression of the acute-phase proteins and inflammatory cytokines.

Any such inflammation leads to local increases in tissue temperature, which may in turn trigger the so-called "heat shock response", which involves the rapid induction of heat shock proteins. Heat shock proteins can also be triggered by other chemical and physiological stresses.

The response is regulated at the level of transcription by a transcription factor referred to as heat shock factor (HSF) in eukaryotes. Unsurprisingly, this transcription factor binds to elements in the promoters of heat shock genes. Arachidonic acid is involved and can induce heat shock gene transcription, as a result of phosphorylation of heat shock factor 1 (HSF1).

HSF1 is normally present in cells in an inactive condition; stress leads to activation, which leads to the induction of trimerisation and high-affinity binding to DNA and to exposure of domains for transcriptional activity.

The heat shock transcription factor (HSF) is the only known sequence-specific, homo*trimeric* DNA-binding protein

The heat stress signal is thought to be transduced to HSF by changes in the physical environment, in the activity of HSF-modifying enzymes, or by changes in the intracellular level of heat shock proteins.

In addition to their role in the heat shock response, many of the heat shock proteins are developmentally regulated; some are believed to act as molecular chaperones, assisting in protein folding.

6.1 Heat shock proteins in *Drosophila*

Heat shock proteins (HSPs) are strongly conserved through evolution in organisms as distinct as bacteria and humans. In common with many areas of molecular biology, this phenomena has been heavily studied in *Drosophila*, the fruit fly.

The three-dimensional structure of the DNA-binding domain of the *Drosophila* heat shock transcription factor has been studied and is similar to the helix-turn-helix proteins discussed earlier in chapters 2 and 3. The DNA-binding structure consists of a four-stranded anti-parallel β-sheet packed against a three-helix bundle.

HSF's transcriptional activating role is believed to be mediated by the carboxyl-terminal region of the protein. Furthermore, deletion analysis and the construction of chimeric proteins have narrowed the key sequence to a region of about 80 amino acids. In both the *Drosophila* and human proteins, this domain is rich in hydrophobic and acidic residues.

The heat shock transcription factor (HSF) is the only known sequence-specific, homo*trimeric* DNA-binding protein. The ability to form trimers depends on the presence of three contiguous GAA boxes present in inverted repeats. It is believed that the trimerisation is mediated via a leucine zipper structure, possibly with the leucine zippers forming a three-stranded coiled coil.

In unstressed cells, HSF1 is found in a complex of approximately 200 kDa and is unable to bind DNA. This complex is believed to involve the protein hsp70. Exposure to elevated temperatures causes an increase in the apparent molecular weight to about 700 kDa, and the protein is then able to bind DNA. Cross-linking experiments suggest that this is a result of the trimerisation of HSF1.

Additionally, inactive, monomeric HSF1 is cytoplasmic, whereas the activated trimer is translocated to the nucleus. Trimeric HSF1 binds with high affinity to heat shock elements (HSEs) in heat shock gene promoters.

6.2 Heat shock proteins in mammalian cells

Mammalian cells express two different HSF1 mRNAs that arise as a result of alternative splicing from a single HSF1 gene. The two HSF1 mRNAs differ by the presence or absence of a 66 bp exon, which is included in the HSF1-α mRNA, but spliced out of the HSF1-β-mRNA. This 22 amino acid sequence is located next to a carboxyl-terminal leucine zipper motif, which is believed to be involved in maintaining HSF1 in the non-DNA-binding monomeric form.

The heat shock element (HSE) in the human HSP70 HSE has been extensively characterised. It comprises three perfect 5′-NGAAN sites and two imperfect sites arranged in tandem inverted repeats. From sequence analysis, some researchers have suggested that the consensus HSE sequence should be 5′-AGAAN. From footprinting experiments, HSF1 appears to bind to all five sites when the gene is being transcribed.

Summary

This chapter introduced a series of transcriptional systems–prokaryotic and eukaryotic–that illustrate the points made in earlier chapters. The tryptophan operon is the paradigmatic anabolic operon and should be contrasted with the lactose operon in chapter 9, the paradigmatic catabolic operon. The β-globin locus control region (LCR) is essentially an enhancer, but it has many properties, including conferring susceptibility to nucleases by altering chromatin structure. Phosphate metabolism in yeast is under transcriptional control and is reminiscent of the prokaryotic systems. The stringent response in bacteria is a major shift in metabolism brought on in conditions of amino acid starvation; it is mediated at the level of transcription. The heat shock response is strongly conserved from *Drosophila* to humans and is also a result of transcriptional changes.

Further reading

1. Sen, A. K. and Liu, W.-M. (1990) Dynamic analysis of genetic control and regulation of amino acid synthesis: the tryptophan operon in *Escherichia coli*. *Biotechnol. Bioeng.* **35**, 185–194.
2. Lenburg, M. E. and O'shea, E. K. (1996) Signalling phosphate starvation. *Trends Biochem. Sci.* **21**, 383–387.
3. Wu, C. (1995) Heat shock transcription factors: structure and regulation. *Annu. Rev. Cell Dev. Biol.* **11**, 441–469.
4. Roesser, J. R., Nakamura, Y. and Yanofsky, C. (1989) Regulation of basal level expression of the tryptophan operon *of Escherichia coli*. *J. Biol. Chem.* **264**, 12284–12288.
5. Dekel-Gorodetsky, L., Schoulaker-Schwarz, R. and Engelberg-Kulka, H. (1986) *Escherichia coli* tryptophan operon directs the *in vivo* synthesis of a leader peptide. *J. Bacteriol.* **165**, 1046–1048.
6. Platt, T. (1981) Termination of transcription and its regulation in the tryptophan operon of *E. coli*. *Cell* **24**, 10–23.
7. Evans, T., Felsenfeld, G. and Reitman, M. (1990) Control of globin gene transcription. *Annu. Rev. Cell Biol.* **6**, 95–124.
8. Ottolenghi, S., Mantovani, R., Nicolis, S., Ronchi, A. and Giglioni, B. (1989) DNA sequences regulating human globin gene transcription in non-deletional hereditary persistence of fetal hemoglobin. *Hemoglobin* **13**, 523–541.
9. Goldman, E. and Jakubowski, H. (1990) Uncharged tRNA, protein synthesis, and the bacterial stringent response. *Mol. Microbiol.* **4**, 2035–2040.
10. Treisman, J., Gönczy, P., Vashishtha, M., Harris, E. and Desplan, C. (1989) A single amino acid can determine the DNA binding specificity of homeodomain proteins. *Cell* **59**, 553–562.

End of unit questions

1. Explain the role of hydrogen bonding in the regulation of the *E. coli* tryptophan operon.

2. What occurs in the bacterium *E. coli* when it has consumed the amino acids available in the medium?
3. Discuss the heat shock response and the role of transcription factors in it.

PROBLEMS

1. Discuss the evidence for our current understanding of the regulation of the tryptophan operon in *E. coli* (see references 1, 4, and 5).
2. Discuss the role of the locus control region (LCR) in the regulation of the human β-globin gene (see reference 7).

Glossary

Activator	A protein that interacts with a DNA sequence or another protein to increase transcriptional efficiency from a gene or genes.
Adenine	A purine base contained within DNA and RNA. Forms a base pair with Thymine.
Alternative splicing	Certain eukaryotic genes can be spliced in more than one way, forming the potential for at least two different proteins. This can be achieved by removing different intron sequences.
Bacteriophage	A virus that can infect a bacterium. Two common bacteriophage used in the laboratory are λ (lambda) and M13.
Basal transcription machinery	The smallest protein complex needed for transcription to occur.
Base	In the molecular biological sense, a base refers to adenine, guanine, thymine, cytosine, and uracil, the components of DNA and RNA.
Base pair	The fundamental unit of double-stranded DNA. The base Adenine pairs with Thymine and Guanine pairs with Cytosine by hydrogen bonding.
Central dogma	The idea that information passes from DNA to RNA to protein.
Chimeric	The artificial joining of two protein domains into one protein, using recombinant DNA techniques.
Chromatin	The complex of DNA with histones and non-histone proteins, found in the eukaryotic nucleus.
Chromosome	The DNA contained within a cell is present in the form of chromosomes, giant DNA molecules containing many genes. Prokaryotic cells typically have one circular DNA chromosome. Eukaryotic cells have linear chromosome(s).
Cis	In DNA terms, *cis* means located on the same, physically contiguous DNA strand within the double helix.
Coding sequence	The DNA strand which is the same sequence as the RNA being formed (except that where a Thymine is present in the DNA, Uracil will be present in the formed RNA).
Co-factor	Additional proteins that interact with the basal transcription machinery.
Conformation	The three-dimensional shape of a molecule.
Consensus sequence	The "average" or most common form of a particular DNA element. Often no gene actually contains the consensus, but some variant of it.
Cytosine	A pyrimidine base contained within DNA and RNA. Forms a base pair with Guanine.

Dimerisation	A term used to describe two proteins interacting with each other. A homodimer is formed if the two molecules are the same, and a heterodimer if they are different.
DNA	The molecule that stores and transmits genetic information. It is composed of two strands running anti-parallel to each other. The two strands are twisted into a double helical conformation.
DNA elements	Conserved short DNA sequences that have a specific function because of their sequence. Typically, a particular protein will bind to these regions.
DNA footprinting	A laboratory technique used to study where and how proteins and other ligands bind to DNA.
DNA polymerase	An enzyme that makes DNA from deoxynucleotide triphosphates using a DNA template; that is, it "duplicates" DNA by base pairing.
Domain	A distinct region of a protein that is able to function independently. If a domain is removed and placed into a new artificial (chimeric) protein, typically it will retain its previous function (such as DNA binding) and confer it on the chimeric protein.
Double helix	The characteristic conformation of double-stranded DNA.
Downstream	Genes are given a direction from the direction of transcription (5′ to 3′). "Downstream" is in the direction of transcription.
Drosophila melanogaster	The full name of a species of fruit fly that is widely used in genetic research.
Electrophoresis	A method of separating molecules by their movement within an electric field. The medium used is usually an agarose or polyacrylamide gel through which the molecules move.
Enhancers	Short DNA sequences that can increase the frequency of the initiation of transcription in a largely distance- and orientation-independent fashion.
Eukaryote	Cell or organism that possesses a nucleus containing the cell's DNA.
Exon	Part of a eukaryotic gene that is transcribed and translated into a protein.
Extracellular	Outside a cell or organism.
Gel shift assay	A laboratory technique for studying DNA-protein interactions using the mobility of the complex in a gel.
Gene expression	The process by which a gene's coding information is converted into proteins, via RNA.
Genome	The entire DNA content of an organism. Its size is given in base pairs.
Genotype	The unique collection of genes present within an organism.
G-protein linked receptor	A receptor present in the cell membrane which when occupied by its ligand sends a message into the cell via a protein that binds and hydrolyses GTP.
Guanine	A purine base contained within DNA and RNA. Forms a base pair with Cytosine.
Haploid	A eukaryotic cell is haploid when it contains a complement of unpaired chromosomes, as in a gamete.
Heat shock response	An evolutionarily conserved cellular reaction and response to excessive heat. The cell protects itself by activating special

	metabolic pathways and the expression of a special set of genes, the products of which enable the cell to survive in the abnormally hot conditions.
Hepatocyte	A liver cell.
Heteronuclear RNA (hnRNA)	Transcribed RNA in a eukaryotic cell that has not yet been processed (splicing, capping, and tailing) into mature mRNA.
Histones	Very basic proteins that form highly complex structures in eukaryotic cells to enable DNA to be packed very tightly into the nucleus.
Homeodomain	A conserved protein sequence found in the homeobox proteins. The homeodomain is a DNA-binding structure.
Hormone	A biochemical "messenger" molecule by which messages are passed between cells and tissues via the extracellular fluids.
Hydrophobic	Water-hating; a property of some molecules. Such molecules or parts of larger molecules will interact with each other (rather than with water), resulting in a non-aqueous environment locally.
Hydroxyl radical	A reactive species, $OH^{\bullet}$, carrying an unpaired electron. Hydroxyl radicals are generated as a biproduct of oxidative metabolism and can be made experimentally in the laboratory.
Intron	A non-coding sequence present within a eukaryotic gene. Introns are spliced out of hnRNA during processing to form mature mRNA.
Ion channel	A membrane protein that allows passage of certain ions through it.
Leucine zipper	A protein dimerisation motif seen in some transcription factors. Leucine side chains protrude from a helical structure in each monomer and these interact to hold the dimer together.
Major groove	When double-stranded DNA forms a double helix, two grooves are created. Due to the shape of a base pair, the two grooves are not the same width. The larger of the two grooves is termed the major groove.
Major histocompatibility complex	A region of DNA on human chromosome 6 that contains many genes relating to the immune system and immune recognition.
Minor groove	When double-stranded DNA forms a double helix, two grooves are created. Due to the shape of a base pair, the two grooves are not the same width. The smaller of the two grooves is termed the minor groove.
Monocistronic mRNA	In eukaryotes, genes are typically organised individually, that is, transcription of one gene produces one transcript which is then processed to one mRNA from which one protein is made. As a result, the mRNA is termed monocistronic.
mRNA	The mature, single-stranded RNA, which ribosomes can translate into protein.
mRNA processing	In eukaryotes, the process of polyadenylation, splicing, and the addition of a 5′ cap structure, by which hnRNA in converted to mRNA.
Multicellular organism	An organism consisting of more than one cell. Usually this allows the cells within an organism to become specialised.

Non-coding strand	The DNA strand that is complementary to the RNA being transcribed. Often referred to as the template.
Nucleoplasm	Generic term for the material in the nuclear environment.
Nucleosome	The basic unit of chromatin. A nucleosome is formed from eight histone protein molecules and the DNA they interact with. Around the octamer of proteins, the DNA molecule wraps approximately 1.7 times.
Nucleotide	The building blocks of DNA and RNA. Nucleotides consist of a base (adenine, cytosine, guanine, thymine, or uracil), a sugar molecule (ribose or deoxyribose), and a phosphate group.
Nucleus	The "control centre" of a eukaryotic cell. The nucleus is a sub-cellular organelle, bound by a nuclear membrane, containing the genetic material.
Operon	In prokaryotic organisms, genes are often grouped together to allow for co-ordinate expression. An operon is a set of genes under common transcriptional control.
Phenotype	The physical characteristics of a particular cell or organism. Phenotype is dictated by the cell's genotype (the genomic material) and the organism's environment.
Poly(A) tail	Multiple adenine nucleotides added to the 3'-end of eukaryotic mRNA during processing.
Polycistronic	Frequently in bacteria, a single RNA transcript actually contains multiple reading frames, so that multiple proteins can be made from it. The RNA transcript is thus said to be polycistronic.
Polysome	One mRNA transcript molecule being translated by multiple ribosomes at the same time.
Prokaryote	An organism, the cell(s) of which does not have a distinct nucleus containing the genetic material.
Promoter	Specific DNA sequences, usually upstream of a gene's transcriptional start site, that allow the formation of a transcription complex and allow the initiation of transcription.
Reporter gene	A gene encoding a protein that has a readily experimentally detectable or quantifiable activity or property.
Repressor	A protein that binds to a DNA sequence or another protein to inhibit transcription from a specific gene.
Ribosomal RNA	RNA present within the structure of a ribosome.
RNA polymerase	The enzyme that transcribes RNA from a DNA template by base pairing.
Sigma factor	A subunit of bacterial RNA polymerases needed for transcriptional initiation.
Single copy gene	A gene present only once in an organism's genome.
Spliceosome	A ribonucleoprotein complex where splicing occurs.
Splicing	Process in which non-coding intronic sequence is removed from primary eukaryotic transcript.
Super shift assay	Experimental technique to study proteins binding to DNA.
TATA box	A DNA sequence found within prokaryotic promoters, typically positioned at approximately position –25.
Thymine	A pyrimidine base contained within DNA and RNA. Forms a base pair with Adenine.
Trans	In DNA terms, *trans* means on a separate DNA molecule.

Transcription	The process by which RNA is produced from DNA.
Transcription factor	A protein, the function of which is to control, positively or negatively, part of the transcription process.
Transcriptional machinery	The enzymes and proteins involved in transcription.
Transfer RNA	A small RNA that serves as a carrier for amino acids and is used by the ribosome in translation.
Translation	Process of producing proteins from mRNA.
Upstream	Genes have a "direction"; transcription proceeds from 5′ to 3′. "Upstream" is in the opposite direction to that of transcription.
Uracil	A pyrimidine base contained in RNA.
Zinc finger protein	The "zinc finger" is a protein motif found in many transcription factors. It is believed that the proteins plus a zinc ion form a DNA-binding structure.

End of Unit Questions and Answers

Chapter 1

1. **List the ways in which an extracellular hormone signal that interacts with a cell-surface receptor can lead to alterations in gene expression.**

   There are several fundamental mechanisms by which an extracellular hormone signal from a cell-surface receptor can be transduced into the cell to alter the rate of transcription. These include (1) receptors operating via G-proteins, (2) receptors that are ion channels, and (3) cytokine receptors that possess cytoplasmic tyrosine kinase activity.

2. **Explain how a steroid hormone affects gene expression.**

   Hormones are extracellular signalling molecules secreted by a cell to cause an effect at another location. The hormone travels to its point of action in the bodily fluids. Steroid hormones diffuse through the cell membrane and into the cytoplasm. There, they bind receptor proteins, activating the receptor. The activated cytoplasmic receptor-hormone complex is then translocated to the cell nucleus, where it binds genomic DNA and alters gene transcription.

3. **Explain two generic mechanisms by which a negative transcription factor might operate.**

   A negative transcription factor can act either by interacting with a positive transcription factor, stopping its action (indirect repression) or by directly interacting with the basal transcription complex of RNA polymerase and associated factors (direct repression).

4. **Why is transcription regulated primarily at the point of initiation?**

   In a word, "economy". Initiation of transcription is the earliest point in the gene expression process. It would be a waste of energy and chemicals to regulate protein production at the point of translation, because by that time the cell would have unnecessarily produced mRNA.

CHAPTER 2

1. **List the post-transcriptional processing steps that eukaryotic RNA transcripts may undergo before the mRNA leaves the nucleus for translation.**

   Within eukaryotic cells hnRNA goes through several stages of processing before it becomes mRNA and is translated into a protein. The key eukaryotic hnRNA processing events are the G- capping of the 5′-end of pre-mRNA, addition of the poly(A) tail to the 3′-end, and the splicing of introns. These processes occur in the nucleus, co-transcriptionally or immediately post-transcriptionally.

2. **What does the "GU-AG" rule refer to?**

   Analysis of many intron/exon boundary sequences has produced the "GU-AG rule". This rule states that the sequence 5′-GU always occurs at the 5′-end of the intron, creating the "donor" site, and the sequence 5′-AG always occurs at 3′-end of the intron, creating the acceptor site. Hence the "GU-AG" rule.

3. **How is it known that DNA binding is mediated primarily by distinct domains within transcription factors?**

   Many mammalian transcription factors have been studied and their DNA-binding properties have been localised to relatively small sub-regions or domains within the proteins. These regions have been strongly conserved through evolution and are found in many DNA-binding proteins. Using molecular biology techniques these domains can be moved and joined onto irrelevant proteins. The resulting chimeric protein typically inherits the DNA-binding characteristics of the protein from which the DNA-binding domain was taken.

CHAPTER 3

1. **Explain the significance of a defect in a transcription factor, compared to–for example–a muscle structural protein.**

   If the gene encoding a muscle structural protein contains a mutation then a defective protein will be produced that will affect the muscle cells. However, it is unlikely to cause problems beyond that. In contrast, a mutation in a transcription factor that itself controls the expression of dozens of other genes will have wide-ranging effects; it will at the minimum affect all its target genes.

2. **What is a homeotic transformation? List two that have been characterised in *Drosophila*.**

A developmental abnormality in which one part of the body develops in the likeness of another is referred to as a "homeotic transformation".

In *Drosophila* the paradigmatic homeotic genes are *Antennapedia,* mutations in which cause the transformation of an antenna to an extra leg and *bithorax,* mutations in which cause the transformation of the halteres (balancing organs) into a second pair of wings. It is now known that both genes, *Antennapedia* and *bithorax,* encode homeobox transcription factors.

3. **How might a chromosomal translocation lead to aberrant expression of a transcription factor?**

   Suppose the chromosome breaks occur in the promoter region of a transcription factor gene and the promoter of a muscle protein. In the translocation the transcription factor gene becomes joined to the muscle protein promoter. In this resultant chimeric gene, the transcription factor will be aberrantly expressed in muscle tissue and may inappropriately turn on expression of its target genes in muscle cells.

4. **What is the homeobox?**

   The homeobox is a 183-base-pair DNA sequence encoding the homeodomain, a 61-amino-acid DNA-binding domain. Genes containing this sequence are commonly referred to as homeobox genes and the proteins they encode as homeodomain proteins. Sequence identity between mammalian and insect homeoboxes is typically 65–80% at the nucleotide level. Homeoboxes exist at least as far back in evolution as the Hydrozoans and they have been found in plants too.

Chapter 4

1. **Explain why the minimum size of a promoter in *E. coli* is 12 bases, given that the *E. coli* genome is 4.6×10^6 bases. What is the minimum size of a human promoter, given a genome size of 3×10^9 bases, and why?**

   For a specific gene to be controllable within a cell it needs to have a unique promoter sequence. In *E. coli,* a 12-base promoter is necessary to ensure the sequence does not occur by chance in the genome. The *E. coli* genome comprises 4.6×10^6 bases, and 4^{11} = 4,194,304; thus, an 11-base sequence could occur in the *E. coli* genome of 4.6×10^6 bases *by chance* alone. Therefore the promoter sequence, to be unique, must be 12 bases or longer as 4^{12} = 16,777,216. A 12-base sequence is unlikely to occur by chance in a genome of that size.

   By the same argument, the minimum length of a human promoter must be 17 bases because $4^{16} = 4.29 \times 10^9$ and the genome is approximately 3×10^9 bases. Extending the promoter by one base, $4^{17} = 1.72 \times 10^{10}$, giving a sequence that is unlikely to occur by chance.

2. **If many transcription factors in a protein family share a conserved DNA-binding domain, how does each bind to different promoters?**

   Given such observations, the conclusion must be that much of the DNA-binding specificity lies outside the actual DNA-binding domain, where the proteins in transcription factor families are much less similar.

3. **How–at the level of transcription–might a cancerous cell "escape" detection by T-cells?**

   T-cells "inspect" peptides presented in MHC molecules and kill cells that present foreign or defective peptides. Thus, a way around this defence mechanism is for a cancer cell not to express MHC molecules. Such a cell would go undetected by T-cells.

   Studies on class I-negative tumour cells have revealed mutations in the promoter region of the class I MHC genes, illustrating that control is at the level of transcription. Down-regulation or complete, sometimes selective, loss of class I antigen expression has been reported in various tumour cells; this down-regulation can be locus- or allele-specific. In some cases this down-regulation or loss is associated with altered levels of the transcription factors that bind these regulatory elements.

CHAPTER 5

1. **Which subunits are necessary for a transcriptionally active bacterial RNA polymerase and in what order do they assemble?**

   In prokaryotes transcription is performed by a single type of RNA polymerase. The "core" polymerase comprises five polypeptides, two alpha (α) chains, one beta (β), one beta′ (β') and one omega (ω) chain. A sixth subunit, sigma (σ), is also associated with the functional polymerase and its role is to direct transcription at specific promoters. Together, these six subunits comprise the RNA polymerase holoenzyme.

   The formation of active polymerase can be followed *in vitro* and complexes smaller than the complete holoenzyme have been identified. From these studies the following assembly order for the *E. coli* polymerase seems likely:

   $$2\alpha \rightarrow \alpha_2 + \beta \rightarrow \alpha_2\beta + \beta' \rightarrow \alpha_2\beta\beta' \text{ (inactive)} \rightarrow \alpha_2\beta\beta' \text{ (active)}$$

   In the final step, a conformational change occurs and the complex is then active.

2. **Why does the ρ protein not cause premature termination of all transcripts?**

   There are two reasons why the ρ protein does not cause premature termination of all transcripts. Firstly, some genes do not require ρ for termination. Their termination signal is in the form of a hairpin loop. These genes operate independently of ρ. Secondly, in all genes, ρ competes for the RNA with ribosomes. The ribosomes translating the mRNA block binding of ρ until the ribosome has passed the stop codon on the RNA transcript.

3. **What is a transcription bubble and how might it be studied experimentally?**

   For a RNA polymerase to transcribe RNA from a gene it must first separate (or "melt") the hydrogen-bonded strands of duplex DNA. The region of the DNA that is melted is known as the transcription bubble. This bubble moves along the DNA with the polymerase. The transcription bubble can be studied experimentally using nuclease digestion and footprinting.

CHAPTER 6

1. **List the types of genes transcribed by the three eukaryotic RNA polymerases and describe the common characteristics of the promoters in these types of genes.**

   RNA polymerase I is located in the nucleolus where it transcribes only the tandem array of genes for the ribosomal RNAs (rRNAs). RNA polymerase I promoters are characterised by the absence of a TATA box and the presence of two key sites, a GC-rich region, located at approximately –45 to +20, and a second GC-rich region, referred to as the upstream control element (UCE), typically at approximately –180 to –107.

   mRNA precursors are synthesised by RNA polymerase II, which is also located in the nucleoplasm. This enzyme transcribes all of the cell's complement of mRNA molecules and several small RNA molecules, such as the U1 snRNA used in the splicing apparatus. RNA polymerase II promoters typically contain a TATA box, 15–25 nucleotides upstream of the transcription start site, though TATA-less promoters are known. Additionally, the sequence Py_2CAPy_5 is usually seen at the transcriptional initiation site, the "A" being position +1 or the first base of the transcript. Often, RNA polymerase II promoters also contain a GC-box centred at around –90. The consensus sequence is 5′-GGGCGG and promoters may contain more than one copy. This sequence is a binding site for the transcription factor Sp1 (stimulatory protein 1). The so-called CCAAT box is also a common element in these promoters.

   RNA polymerase III transcribes the 5S rRNA, tRNA, and snRNA genes. It is located in the nucleoplasm. Two types of promoter are known in RNA polymerase III genes; the promoter is downstream of the transcriptional start site in the 5S rRNA and tRNA genes and upstream of it in the snRNA genes. The 5S RNA gene promoter contains no TATA box and is located at +55 from (*i.e., downstream* of) the transcriptional start site (referred to as Box A). A second key sequence is located at +80 to +90 (referred to as Box C). TFIIIA binds the BoxA–BoxC region and appears to promote the binding of TFIIIC to the same area. Similarly, the tRNA genes contain two key sequences, both downstream of the transcription initiation site. Again, there is no TATA box. The second type of RNA polymerase III promoter is seen in the snRNA genes. This is found upstream of the gene and is fairly variable among genes and different species. Such promoters contain binding sites for the OCT transcription factor, a proximal sequence element (PSE), and a TATA box located close to the PSE.

2. **Describe the nature and role of TFIID in eukaryotic transcription.**

TFIID comprises the TATA box binding protein (TBP) and TBP-associated factors (TAF_{II}s). TFIID appears to be the only general transcription factor to have the ability to bind to DNA in a sequence-specific manner. It is also the first subunit of the pre-initiation complex to associate with the DNA, where it initiates the recruitment of RNA polymerase II and other general transcription factors.

Chapter 7

1. **Describe the structure of the nucleosome.**

The nucleosome is formed from an octamer of the histone proteins, H2A, H2B H3, and H4. This octamer is built from a heterotetramer of two H3 and two H4 units, to which two heterodimers of H2A and H2B are joined. DNA is wrapped around this octamer complex approximately 1.7 times.

2. **What happens to the nucleosome during transcription?**

In short, it remains unknown exactly what happens to the nucleosome. It is clear that transcription occurs on DNA associated with nucleosomes. If co-occupancy of nucleosome cores and transcription factors can occur, it may well be that transcription occurs in the presence of histones. The retention of histones on DNA may be incompatible with the binding and engaging of RNA polymerase. If so, there must be a mechanism for removing nucleosomes from promoter elements either by removing the nucleosome cores from the DNA or by moving the nucleosome cores along the DNA. Two models for nucleosome movement on DNA to allow transcription have been proposed. The simplest mechanism of clearing nucleosomes from the promoter and enhancer regions would be nucleosome "sliding", in which nucleosomes displace from transcription recognition sites in response to other proteins binding. A second possible method of generating a nucleosome-free region to enable transcription factor binding would be simple dissociation of nucleosomes from DNA. While transcription factor binding to nucleosomal DNA does not directly result in nucleosome displacement, factor binding can allow the movement of nucleosomes onto other DNA regions (either *cis* or *trans*).

3. **Discuss the role of $TAF_{II}250$ in eukaryotic transcription.**

   The 250 kDa RNA polymerase II TATA-binding protein (TBP)-associated factor ($TAF_{II}250$) is the core subunit of TFIID and interacts with a variety of other TAFs as well as TBP. $TAF_{II}250$ is necessary for the activation of particular genes and associates with components of the basal transcription machinery, including TFIIA, TFIIE, and TFIIF. In addition, $TAF_{II}250$ has both kinase and histone acetyltransferase functions.

CHAPTER 8

1. **Discuss the binding of the λ repressor to the operator sequence.**

   The λ repressor is capable of binding to all three operator regions within the *cro* and *repressor* gene promoters; however, the affinity of repressor for each site is different. Specifically, the repressor has the highest affinity for O_R1, approximately ten times higher than that for O_R2 or O_R3. For activation of the P_{RM} promoter and hence production of more repressor, the repressor needs to bind to both O_R1 and O_R2. Logically then, there is a problem; if there is a ten-fold difference in affinity between O_R1 and O_R2, there would need to be a ten-fold increase in concentration of repressor to allow its own production. However, as the P_{RM} promoter is inactive there can be little increase in repressor concentration. The phage overcomes this potentially fatal problem by using interactions between repressor dimers; when the first repressor binds to O_R1, it immediately causes a change in the DNA structure such that further λ repressor binds to O_R2 with higher affinity. The second λ repressor dimer not only binds O_R2, but also interacts with the repressor dimer bound to O_R1. As a result of this interaction, the O_R1 and O_R2 sites fill almost simultaneously despite the *apparent* ten-fold difference in affinity.

2. **How does the phage decide between the lytic and lysogenic pathways?**

   The decision between the lytic and lysogenic pathways is made by the amount of the protein cII. If production of cII is high after infection, then the phage will enter the lysogenic life cycle, whereas if cII activity is low, the lytic life cycle will be followed. The cII protein is relatively unstable, and is susceptible to protease digestion. Many proteases are present in the host cell and environmental conditions, such as the availability of nutrients, will affect their concentration. If the host cells are in a nutrient-rich environment, proteases will be active and the phage is more likely to enter the lytic life cycle. If the host cells are growing in a nutrient-poor medium, protease activity will be lower and the phage is likely to enter the lysogenic life cycle.

3. **What role does the malB protein play in λ infection?**

   The malB protein is present on the cell surface of the *E. coli* bacteria. It is used by the phage for cell attachment; it acts as a "λ receptor".

CHAPTER 9

1. **How did the super repressor, I^s, mutants further our understanding of the operation of the lactose operon?**

   I^s mutants provided important evidence that the control effected by the *I* gene (*lac* repressor) was mediated by a cytoplasmic component (of course, with hindsight, a transcription factor). I^S "super" repressor mutants have completely lost their ability to synthesis β-galactosidase, permease, or transacetylase. Furthermore, it can easily be shown that this is not simply the results of deletion of these genes or their promoter. Again with hindsight, the most consistent way of explaining this type of mutant would be to posit that the *lac* repressor expressed in such bacteria can still associate with its binding site as tightly as the wild-type repressor, but that it has lost its ability to be inactivated by lactose. Presumably, in the I^S mutant, the mutations are such as to alter the binding site for lactose so that it no longer binds or prevent the conformational change when it does bind.

2. **What is the role of β-galactosidase in lactose metabolism?**

   The enzyme β-galactosidase metabolises lactose into glucose and galactose.

3. **Explain equilibrium dialysis and its use in the study of the lactose operon.**

   Equilibrium dialysis is a technique that can be used to characterise proteins. A dialysis membrane allows the passage of molecules up to a certain size; larger molecules cannot pass through it. If a freely diffusible (small) molecule starts to concentrate onto one side of the membrane then it must be binding to a particular molecule present on that side of the membrane that cannot pass through it (*i.e.*, it is too large).

   This technique was used to isolate the *lac* repressor. The dialysis experiments were performed using a radiolabelled mimic of lactose, ^{14}C-IPTG, which binds to the *lac* repressor but cannot be metabolised by β-galactosidase. By performing a series of experiments using different fractions from *E. coli* preparations, it was possible to determine which fraction concentrated the ^{14}C-IPTG and therefore contained the *lac* repressor molecule.

CHAPTER 10

1. **Explain the role of hydrogen bonding in the regulation of the *E. coli* tryptophan operon.**

   The tryptophan operon is regulated firstly by a repressor protein (similar to that in the lactose operon) and secondly by a mechanism known as attenuation, which relies on hydrogen bonding. Attenuation uses hairpin loops that form in the mRNA. If a certain hairpin loop forms, translation will be terminated. In the presence of tryptophan, such termination occurs. In the absence of tryptophan, however, a different hairpin loop forms in the mRNA and translation will be completed.

2. **What occurs in the bacterium *E. coli* when it has consumed the amino acids available in the medium?**

   When *E. coli* is starved of amino acids, a mechanism known as the "stringent response" will be triggered. In the stringent response protein biosynthesis decreases and there is a switch to producing only the enzymes for the amino acid biosynthesis pathways, such as those of the tryptophan operon. The synthesis of all other non-essential biomolecules is greatly reduced, including that of rRNA, tRNA, phospholipids, and carbohydrates.

3. **Discuss the heat shock response and the role of transcription factors in it.**

   The heat shock response is strongly conserved through evolution in organisms as distinct as bacteria and humans. In common with many areas of molecular biology, this phenomenon has been heavily studied in *Drosophila*, the fruit fly.

   The three-dimensional structure of the DNA-binding domain of the *Drosophila* heat shock transcription factor has been studied and is similar to the helix-turn-helix proteins. The heat shock transcription factor (HSF) is the only known sequence-specific, homo*trimeric* DNA-binding protein. The ability to form trimers depends on the presence of three contiguous GAA boxes present in inverted repeats. It is believed that the trimerisation is mediated via a leucine zipper structure, possibly with the leucine zippers forming a three-stranded coiled coil.

   In unstressed cells, HSF1 is found in a complex of approximately 200 kDa and is unable to bind DNA. This complex is believed to involve the protein hsp70. Exposure to elevated temperatures causes an increase in the apparent molecular weight to about 700 kDa, and the protein is then able to bind DNA. Cross-linking experiments suggest that this is a result of the trimerisation of HSF1.

Additionally inactive, monomeric HSF1 is cytoplasmic, whereas the activated trimer is translocated to the nucleus. Trimeric HSF1 binds with high affinity to heat shock elements (HSEs) in heat shock gene promoters.

Index